The Accidental Reef

and Other Ecological Odysseys in the Great Lakes

The Accidental Reef
and Other Ecological Odysseys in the Great Lakes

LYNNE HEASLEY

Illustrations by Glenn Wolff

Foreword by Jerry Dennis

Michigan State University Press ⁓ East Lansing

∞ The paper used in this publication meets the minimum requirements of ANSI/NISO Z39.48-1992 (R 1997) (Permanence of Paper).

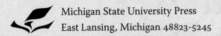 Michigan State University Press
East Lansing, Michigan 48823-5245

Library of Congress Cataloging-in-Publication Data
Names: Heasley, Lynne, author. | Wolff, Glenn, illustrator. | Dennis, Jerry, other.
Title: The accidental reef and other ecological odysseys in the great lakes / Lynne Heasley, illustrations by Glenn Wolff, foreword by Jerry Dennis.
Description: First. | East Lansing : Michigan State University Press, 2021. | Includes bibliographical references and index.
Identifiers: LCCN 2021011126 | ISBN 9781611864076 (paperback) | ISBN 9781609176822 (PDF) | ISBN 9781628954494 (ePUB) | ISBN 9781628964431 (Kindle) Subjects: LCSH: Reef ecology—Great Lakes Region (North America) | Natural resources—Great Lakes Region (North America) | Great Lakes Region (North America)—Environmental conditions. Classification: LCC QH104.5.G7 H44 2021 | DDC 577.7/8909—dc23
LC record available at https://lccn.loc.gov/2021011126

Cover and book design by Erin Kirk
Cover art by Glenn Wolff

ℊ green press INITIATIVE Michigan State University Press is a member of the Green Press Initiative and is committed to developing and encouraging ecologically responsible publishing practices. For more information about the Green Press Initiative and the use of recycled paper in book publishing, please visit www.greenpressinitiative.org.

Visit Michigan State University Press at www.msupress.org

Contents

Foreword *Jerry Dennis*

> All nature is so full, that that district produces the greatest variety
> which is the most examined.
> —Gilbert White, *The Natural History and Antiquities of Selborne*

The fullness of nature—its seemingly endless variety and complexity—is why the eighteenth-century parson and naturalist Gilbert White could spend much of his life studying, keeping records, and writing letters about a small corner of county Hampshire, England. And it's why countless writers and scholars since have done similar work in every corner of the planet. The abundance of the world practically guarantees that when you tug on a thread you'll find that it's attached to a tapestry.

This book by Lynne Heasley joins a long and very rich tradition of writing about nature's tapestry. From the philosophical to the personal, from the lyrical to the scientific, from songs of praise to environmental calls to arms, works of every description, in every language, have attempted to understand the physical world and our place in it. In the process, they have furthered White's goal of paving the way to a "universal correct natural history." Not incidentally, those works say as much about us as they do the world.

Few of us can forget how much the world has changed in the three hundred years since Gilbert White was born. From our perspective in the twenty-first century, deep in the Anthropocene, we see that human impact is everywhere, from the tops of the highest mountains to the bottom of the deepest seas, from the polar ice caps to the great interior deserts, from the atmosphere above us to the aquifers below. The study of nature has never been more important for the future of the planet—and the subject has never been more complex. In an age when natural and human systems overlap, when technology and ecology are becoming more entwined every year, and climate change threatens to bring greater and more destructive changes in the decades ahead, we require a literature that reflects and responds to the complexity of the world we have altered.

Lynne Heasley is eminently qualified to create such a literature. Her knowledge of history and science, her impeccable scholarship, her perceptiveness, disciplined field work, and superb literary skill are tools she uses to synthesize and cross-pollinate ideas across a spectrum of disciplines and produce a deeply textured portrait of place. She is one of the few writers I know with the skill to make the "traumatic wave crests" of industrial chemical pollution fascinating. She can take us from the history of canal-building in Russia to the history of taxonomy, while never losing sight of her primary subject, a section of the St. Clair River that forms the border between Michigan and Ontario. She creates composites, assembles inventories and catalogs, upends conventional research methodologies—all for the purpose of informing us, exciting our interest, making us see the world differently, and inspiring us to care about places that are in danger of being lost.

Scratch the surface anywhere and you'll uncover an inexhaustible complexity. It's no wonder a woodlot or a city block or a stretch of river can engage a writer, an artist, or a scientist for a lifetime. Eudora Welty, who spent most of her life writing brilliant and enduring stories about a small town in Mississippi, wrote, "One place understood helps us understand all places better." The only way I know to truly understand a place is to become thoroughly immersed in it, exploring it in as many ways and from as many points of view as possible. In the pages that follow, Lynne Heasley takes us to the cutting edge of ecological study in places that have been historically and lamentably overlooked. I hope you'll find her odysseys as fascinating as I have, and that you come away from this book with your appreciation for the Great Lakes region—and with your curiosity and sense of wonder—enhanced and stimulated. It's a big world out there. How wonderful that getting to know its hidden corners can make it even bigger. I'm convinced that the future of the planet depends upon such knowledge.

Preface

The Accidental Reef and Other Ecological Odysseys in the Great Lakes rotates natural history, historical, biographical, and experimental chapters to form a kaleidoscope of the Laurentian Great Lakes. Our first prism is the natural world below the surface of the St. Clair River, whose impact radiates throughout the basin. The reef itself is the volume's center axis, or homeplace.

From the nineteenth to early twentieth centuries, the Great Lakes–St. Lawrence region was the industrial epicenter of the United States and Canada. Paper, chemical, auto, steel, and other industries defined and became defined by the region. During this time, lake sturgeon nearly went extinct from overfishing, industrial pollution, and engineering works that eliminated habitat and disrupted seasonal migrations to make way for a transoceanic shipping waterway.

Within a broad-scale Great Lakes picture, the Huron–Erie corridor—which is the hydrologically unified and undammed St. Clair River–Lake St. Clair–Detroit River system—was practically the waterway capital of North American industrial development. For example, huge salt and brine deposits laid the foundation for the rise of the chemical industry and Dow Chemical; steamships traveled along the corridor, and near a salt mine at Algonac, ships dumped their coal waste (coal cinders or "clinkers"). This happened over several decades and in the same place. Unbeknownst to anyone, a modest bed of coal clinkers became a spawning reef, a biological harbor for lake sturgeon during decades when their populations were in greatest peril, "the place no one knew," to riff on David Brower.

No one knew about this reef. Also, no one knew about a second and still more important spawning reef a few miles upstream on the St. Clair River, near the Blue Water Bridge. The story of their discovery is woven through the second part of this volume, "On Seeing and Knowing: An Underwater Biography," which emerged from my series of conversations with diver-filmmaker-artist team Gregory A. D., Greg Lashbrook and Kathy Johnson.

One core argument of *The Accidental Reef* is that care and protection of the Great Lakes demand different ways of seeing and knowing. Notes in the volume reinforce the argument. They are a parallel narrative, another meeting ground where natural scientists, humanists, artists, specialists, practitioners, and local and Indigenous communities can see themselves as interrelated and in dialogue. Explore the notes to discover who is doing what, or to follow a literary path.

Also woven throughout is an exploration of abundance: what it means for the natural world, and for the Great Lakes in particular—for their care, conservation, sustenance—that the place holds cascading opportunities for extraction and pollution at levels beyond comprehension, and at accelerated rates that oftentimes seem to foreclose more sustainable alternatives.

Finally, the reef itself is a quintessential Great Lakes site, one moment complex and mercurial, another moment conducive to meditation. This accident of history and ecology is a perfect underwater marina from which to set forth and explore the Great Lakes, and to search for such safe harbors as restoration and sustainability. Most importantly, this small-scaled, barely known place is awe-inspiring on grand scales of time and space, and humbling for its cosmic reminder that chance matters.

Part I. Freshwater Reef

A WORLD BELOW AND BEYOND

COAL CLINKER

SCUTE

Lake Huron

MICHIGAN

St. Clair River

Algonac

ONTARIO

Detroit River

Lake St. Clair

Windsor

Acipenser fulvescens

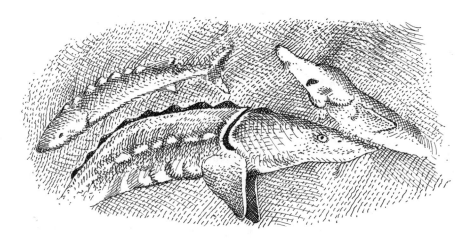

At the Reef

By the banks of the St. Clair River, twenty feet deep on the North Channel floor, lies an accidental reef. Cobbles make up the reef, each gray or red-tinged rock burnt to its present form. Manganese, iron, feldspar, quartz: fused in fire, like lava. Melted minerals. Cooled, broken chunks of aluminosilicate. In the vernacular, coal clinkers. The reef is mere ship waste, stripped of its combustible carbon over a century ago, then strewn and sunk with no particular design or function. Powerful river currents have arranged the remnant cinders into a more orderly matrix of matter and small spaces. The reef could fill an Olympic-sized pool, miniscule on a Great Lakes scale. Yet this reef is a landmark on the bottom of the inland seas, a beacon to underwater travelers.

Sometime in late May, the water temperature in the North Channel will reach 56 degrees Fahrenheit. This is a biological signal. At 56 degrees, freshwater giants wintering in Lake St. Clair, or at the outlet from Lake Huron farther north on the river, will begin their seasonal transhumance. They look like prehistoric submarines, each camouflage-gray body tapered at the snout, sleek scaleless skin, unevenly divided tail like a shark's, armored scutes running gill to tail fin—five jagged rows, a barb jutting from each scute plate.[1] The fish will congregate in a staging area less than a mile upstream from the reef.[2] They'll jostle and thrash and porpoise at the surface. They'll explore the bottomland, scraping the substrate with four whiskery organs called barbels that dangle from their snouts. Imagine the tactile sensation of running your fingers along a rough brick wall, only with fingers that also have taste buds.

Taking turns, a threesome or more will cruise to the rocky reef—an ovulating female and her male escorts. Fish in heat. A *ménage-à-poissons*. The female might be seven feet long, over two hundred pounds. Upwards of 500,000 eggs wait inside her, adding sixty pounds to her girth. She could be over one hundred years old, or as young as twenty-five. Her attendants are smaller, and will die younger.

At the reef, the males make a spectacle of themselves. First they joust, to get the best position by their female. *Skull fractures have been observed*, note two fish biologists.[3] (Are there times when scientific detachment seems forced?) Then the males pound the female's egg-swollen belly, four tail-thumps a second. They grunt to the beat, a wump-wump-wump-wump in cadence with the tail and caudal peduncle. Finally, they ejaculate. *Recuperating between bouts*, say our scientists, *some males will simply lie against a rock or log, or even on the legs of the observer.*[4] Spockish humor again? One eyebrow raised while writing for the journal, *Applied Ichthyology*? Who wouldn't picture these flopped menfish indulging in a postcoital cigarette, completely spent after four seconds? But in sturgeon time, these four seconds—whose final victory, a Milky Way of sperm expanding across the watery sky above the reef—have thrummed their seasonal drumbeat for 150 million years.

For her part in the ancient ritual, the female sturgeon might first nuzzle and prod loose pebbles with her beak, sweep and fluff soft silt with her flicking tail. Preparations for an aquatic nursery. Then she wiggles her swollen body while the males pound, until she expels a black trail of eggs from her dilated, purpled vent (or urogenital pore). The trail scatters in an instant from fast currents and the swirls of water around the thrashing fish. Lake sturgeon are broadcast spawners. They release their sperm and eggs in open water, to fertilize and be fertilized in chance parental encounters. But not all the eggs at once. It will take four, six, maybe eight hours. A half-million eggs later—a whole universe of potential life released from her body—the female will look gaunt, and exhausted. Her belly will have stretch marks. She and her sister Methuselahs at the reef will not spawn again for at least four years.

The spawning female is abundance personified: massive fatty flesh, eggs (or roe) in fruitful numbers to conjure a pantheon of goddesses—Abundantia, Copia, Feronia, Fortuna. In abundance lies her kind's greatest risk . . .

Each unmoored egg is about four millimeters, the size of a peppercorn. Despite their smallness, the eggs are not newly formed. Journey from ovary to reef is years long. Biological processes with neo-Latin names make up the egg's timeline, or lifeline, inside its mother: oogenesis, meiosis, vitellogenesis. These encompass the egg's development from stem cell to immature egg cell to mature egg with yolk. Hormones trigger each stage until the female ovulates, breaking the bond between egg and ovary, freeing it into her body cavity, through the oviduct, and out to its fate in the St. Clair River. The egg will be three or four years old. For comparison, a hen makes and releases an egg in twenty-six hours.

Sturgeon eggs pepper the water, descending through male milt. Something special happens when a sturgeon egg meets water. Apparently, "a sialic acid mediated chemical reaction hydrolyzes [protein] compounds on the egg surface creating an adhesive property."[5] In other words, the egg secretes a natural glue. The upshot: Eggs will stick themselves to coal cinders or any other substrate. Each makes landing and settles fertilized. The reef itself is a massive pile of rock two meters deep (6.5 feet). The structure is porous, with small spaces sized perfectly to harbor sturgeon sac fry when they hatch. In days the sticky egg coat will rupture (meantime the larva within begins to consume its yolk and grow). Unmoored once more, the egg drifts deeper into the reef, away from predators.

Lake sturgeon. *Acipenser fulvescens*. The species is prehistoric, a "living fossil"; it swam and swept and thumped and spawned when dinosaurs ruled.[6] During the Pleistocene, the species hunkered down in two outpost refuges while the Laurentide Ice Sheet heaved its mass through time. This frozen tidal wave overwhelmed rivers and lakes and biota in its path. It locked up 20 million cubic kilometers of the earth's water—a quantity whose precise volume numbs the mind into abstraction. But ice two miles thick, whose weight compressed the earth's crust three hundred meters like a foot might compress foam rubber a few inches—this is easier to imagine.[7] In glacial history, time and space are inseparable. Ice re-sculpted the first great lakes and rivers of the region into today's Great Lakes–St. Lawrence basin. And here, lake sturgeon reign as North America's largest freshwater fish.[8] The largest fish in the largest system of freshwater in the world. But an awesome size, and a lineage across eons, and a home in our inland seas are not talismans; they do not promise safe passage to the future.

Once, a steamship served a salt mine on the shores of Lake St. Clair, at Algonac. The mine was just one port of entry to vast Michigan brine deposits. This was the rich (not barren) salted ground on which to build an industry in modern chemicals. Herbert Henry Dow, the future face of the industry, chose this salt-fertile country well. But long before Dow Chemical was a household name (before Styrofoam or saran wrap, napalm or Agent Orange, or Bhopal), a steamship's fireman hacked at the burnt buildup lining the coal furnace like a thick coat of lava paint, and dumped the pieces overboard. His railroad brethren did the same; you can walk along train tracks today and find small bits or even large chunks of some long-ago fireman's face-scorching, lung-blacking work. Furnaces were less efficient then, and the coal less pure in carbon. And so, the sturgeon's intricate biological drama collides with other dramas—historical, ecological, geological. This is what we must strain to see on the bottom landscapes of the Great Lakes, out of our everyday sight, these collisions and chance encounters through time.

From "Egg Extraction," an unintended poem in two sturgeon stanzas:

Operation requires 4 people: 2 to strip the eggs and 2 to prepare the eggs for incubation.[9]

When Female is Sacrificed:
Put fish in tube net.
Kill by a blow to the head.
Hang fish by gill or mouth.
Cut off tail and bleed.
Put bowl under fish.
Cut the abdomen open.
Eggs will fall from belly into bowl.
Use fingers to scrape out remaining eggs.

When Female is Not Sacrificed:
Preferred when working with endangered species.
Fish put in stretcher with water tube apparatus attached.
A 8–10 cm incision is made in the abdomen.
Eggs are removed with spoon.
Incision is sutured and disinfected.

Underwater Rashomon

One would think that hundreds of thousands of hatching sturgeon fry would, for their moment, be the dominant life on one small reef in one river channel. Not so. They're not the sole residents of these old coal clinkers. Two-inch Vikings from the Old World found the same rocky crags an ideal outpost. Zebra mussels and round gobies: fierce ecological warriors, they sailed a blink ago from another ancient sea.

L. P. Hartley once wrote that the past is a foreign country.[1] Aquatic residents of today's St. Clair River would not recognize the same river of 1987. This was the year before zebra mussels. The precise route these voyageur mussels traveled is unknowable.[2] Their likely journey started circa 1988, when an ocean carrier unloaded its cargo in northern Europe. Perhaps the ship made stops in the Baltic Sea, and then entered the enormous Port of Rotterdam in the Netherlands. Rotterdam is Europe's maritime hub; in 1988 Rotterdam claimed to be "the busiest port in the world." No one knows the ship or place for sure. There could have been more than one ship, and more than one place. The key is a stop in fresh or brackish water, because zebra mussels are freshwater animals. The empty ship would need weight for stability. So its crew pumped nearly a million gallons of seawater into a system of ballast tanks on the bottom and sides of the vessel.[3] Swept up with the water were free-floating veligers: zebra mussel larvae, 0.1 millimeter around, the same as a human hair, with baby shells visible only under a microscope. Ship ballast holds became their swimming pools. An international trade route became an invasion corridor.

Over weeks the veligers grew as their mother ship crossed the Atlantic Ocean to the St. Lawrence River, then traveling upriver from Montreal to Lake Ontario, then Great Lake to Great Lake. The mouth of the St. Clair River was an important fuel stop mid-route along the St. Lawrence–Great Lakes waterway; it was also a boundary between the Canadian and US cities of Sarnia and Port Huron. In May or June, when the ship's crew refueled or took on new cargo at Sarnia, they off-loaded an equivalent weight of ballast water.[4] Disembarked veligers grown to their juvenile stage settled on a hard surface at the bottom. The rest swam with the currents until they became juveniles and settled. Here they would transform their new home. The past would soon become a foreign country.

On first contact in the Americas, in Lake St. Clair, zebra mussels—*Dreissena polymorpha* of the family Dreissenidae—launched something akin to species-on-species combat. Hundreds- or thousands-to-one, they would descend on some unfortunate native: indigenous clams like *Leptodea fragilis, Pyganodon grandis, Lampsilis siliquoidia, Potamilus alatus, Eliptio dilatata*, and fifteen others of the family Unionidae, or unionids.[5] A military field manual for a mussel attack might read like this:

> Attach yourself to the rear of the first unionid you find. Do this by planting your foot on its shell. Note that your foot is actually a muscle. Press down. This stimulates the byssal gland in your foot to make upwards of six hundred threads. Use the cement-strong grip of your sticky byssal threads to hold fast. (Should you wish to detach, release an enzyme to dissolve the byssal-thread proteins.) Your opponent is large while you are little. With your comrades, pile on. Fifteen thousand dreissenids once sacked a single unionid! So weigh down your clam, disrupt its movement, gash its home ground. Then, as the unionid inhales, creating a micro-current in the water to suck micro-food into its waiting vent, intercept the food; filter it out of the water column. The native's shell thus fouled, it will starve from your "intimate interference."[6]

Warfare is one way to see what happened below the surface of Lake St. Clair. Shell crashed on shell, and eighteen unionid species dwindled then died out. The whole process—invade, colonize, dominate, extirpate—took eight years.[7] This is how a unionid would remember 1988 and beyond, if a clam had memories. Local clams were there first. They didn't deserve their fate. The place is poor for their loss. An ecological yin of interdependence, adaptation, and balance countered by a brutal new yang of conflict, dislocation, and instability. One web of life torn apart, and yet . . .

Pause for a moment on words like "native," "invade," "colonize," "foul," "extirpate," all in the scientific literature on zebra mussels. Are these objective verbs and nouns? Facts on the ground? Exact accounts of ecosystem dynamics? Perhaps they're subjective. Dramas with victims and villains. Moral metaphors. Tragedies. Biology's Latinesque language lends itself to science *and* storytelling, omniscience *and* emotion, Minerva *and* Mars.

So how would *Dreissena polymorpha* retell the story, if mollusks had memories?

Surely a zebra mussel would not begin her story in 1988. She could go back two centuries, when Europe built an industrial system of canals linking its major waterways. Canals are part of her story. Yet two hundred years ago would not be early enough. She would better begin two hundred *million* years ago, in the prehistoric Tethys Sea. But delivering a two-hundred-million-year account of one's kind? What a balance of sweep and detail, of momentum and rhythm. Hers would be an epic of biblical proportions, including the ordeals—floods, plagues, exile. With fair warning, then, to the evolutionary biologists who pioneered the field of systematics and the subfields of taxonomy and phylogenetics and the sub-sub-field of phylogeography, who drew and debated intricate phylogenetic trees, and who mapped the origin story of the dreissenids:

To begin. Like animals, seas and lakes have lineages. The Tethys Sea was the common ancestor of Eurasia's Caspian, Black, and Azov Seas. Tethys divided Laurasia and Gondwana, the northern and southern halves of world continent Pangaea. Gondwana and Laurasia had begun their slow-motion shatter through the Mesozoic Era, planet Earth's middle age. As plates collided, as mountains rose, as continents drifted, as ice sheets advanced and retreated: The Tethys Sea begat the Sarmatian Sea, which begat the Maeotic Sea, which begat Pontic Lake-Sea, which begat Late Pontic Lake-Sea and Babadjan Lake-Sea; and Late-Pontic Lake-Sea begat Kimmearian and Kujal'nik Lake-Seas, while Kimmearian Lake-Sea begat Gurian Lake-Sea, which begat Chauda Lake-Sea, which begat the Ancient Euxinian Sea, which begat the Uzanlar Sea, which begat the Karamgat Sea, which begat the Girkan Sea, which begat Neo-Euxinian Lake, which begat the early Black Sea and the Sea of Azov. Babadjan Sea begat Balakhan Lake, which begat Ackchagyl Lake-Sea, which begat Apsheron Lake-Sea, which begat Baku Lake-Sea, which begat Khazar Lake-Sea, which begat Khvalan Lake-Sea, which begat the Caspian Sea.[8]

And this sea-line from Tethys through eons was water filling earthen goblets—the Euxinian depression, the Pannonian depression, the Caspian depression. The table under the goblets was shaking. So this was water in motion, water in time, water that, with each seismic collision or climatic swing, filled then drained, grew then shrank, connected then quarantined. The goddess Tethys was sister and wife of Oceanus, the World Ocean. "Their child," said the poet Ovid, "couples with sky-lifting Atlas."[9] Ancient Tethys united the Atlantic and Pacific Oceans into a single World Ocean. But the earth moved and Tethys was isolated. Titanic new mountains lifted the sky—Caucuses, Carpathians, Balkans, Alps—and the sea gave way to her smaller descendants.

Tethys was goddess of freshwater. Her brother Oceanus was lord of earth's saltwater seas. In myth their waterbodies comingled. In history the same waters did too. When sea level rises, relative to land, geologists call the event a transgression. When sea level falls, revealing seafloor, geologists call the event a regression. The

Caspian, Euxinian, and Pannonian depressions made up what geographer Andrei Chepalyga called a "cascade system of seas and lakes," a superbasin.[10] (The Great Lakes make up a cascade system, too, a young superbasin.) Every transgression connected the ocean's Mediterranean Sea to the basin, while every regression separated them. Every transgression increased water salinity in the basin, while every regression decreased it. And this is where geology intersected with biology: Every transgression washed Mediterranean sea life into the basin, a healthy home in those eras of new-salted water. While every regression trapped salty species in a slow-building brew of freshwater. Every transgression pushed freshwater life to less-salty margins—brackish bays or upriver. While every regression brought them back to open water. As Tethysian waters dissolved into Caspian, a few freshwater animals evolved to tolerate more salt, while a few saltwater animals evolved to tolerate less. Miocene, Pliocene, Pleistocene, Holocene: In the Ponto-Caspian basin, geology's epochs became biology's epics.

Over millions of years, Nature tested the Dreissenidae for salinity, for temperature, for disturbance. The lineage survived its tests of geology and biology. It became one branch on the tree of life, one parent node and all its offshoots, one common ancestor and all its descendants.[11] Biologists call the evolutionary tree a "phylogeny" and each branch a "clade." You can have a clade within a larger clade within a still larger clade, or a branch of a bigger branch of the biggest branch, until the tree assumes its full shape. You can follow a limb's long sweep. For instance, Animalia and *Dreissena polymorpha* are one beginning and one end along life itself. The nodes between them are trail markers in time; at each fork they signal a way to the present:

> Animalia to Mollusca to Bivalvia to Autobranchia to Heteroconchia to Euheterodonta to Imparidentia to Neoheterodontei to Myida to Dreissenidae to *Dreissena* to *Dreissena polymorpha*.[12]

Or, you can focus on a genealogy—a whole map of family relations, from a node when the ancestral branch splits, through sibling and cousin branches, to their outermost descendants at the tips. Dreissenidae emerged in the Tethys Sea two hundred million years ago. *Dreissena* was one of Dreissenidae's progeny. Sixty-eight million years ago, *Dreissena* split into two offspring, then they both split into two offspring (21 and 17 million years ago), and the four sisters and cousins into five more (from 840,000 to 500,000 years ago), and out to the present tips of the *Dreissena* clade:

> *Dreissena carinata, Dreissena blanci, Dreissena anatolica, Dreissena polymorpha, Dreissena caputlacus, Dreissena rostriformis bugensis.*[13]

Dreissena polymorpha. At 500,000 years old, a twig on the tree of life. A distant heir of *Dreissena* but also a creation of Ponto-Caspian geology. With each upheaval, *Dreissena*'s clade innovated new ways of living.[14] Her descendants perfected the free-swim in their larval stage, which allowed them to travel long distances, or through hostile waters to better territory. *Dreissena polymorpha* became one of the "euryhalines," creatures who could survive shifting levels of salt; it became eurythermal, tolerant of a wide temperature range. The species developed an especially firm grip. It could affix faster and hold tighter to hard surfaces than its relatives.[15] This was helpful in strong water currents. Then those two of nature's primal powers—geology and biology—intersected with human history, a younger force of nature.

And here, canals enter the story.

———

In 1803, John Phillips published *The General History of Inland Navigation; Containing a Complete Account of All the Canals of the United Kingdom with Their Variations and Extensions according to the Amendments Acts of Parliament to June 1803 and a Brief History of the Canals of Foreign Countries.*[16] The book was less a Britannic than a world encyclopedia of canals. A loquacious encyclopedia. "This work," began Phillips, "having passed through three editions in quarto, the Author [Phillips, referring to himself] has been advised, for the purpose of a still more general circulation, to abridge it of such matter as was least useful; which has enabled him [Phillips again] not only to reduce it to a moderate price, but also to add a large quantity of new and important information. He has also given the plan of a lock to save water, which will be found worthy of particular attention in such places as are liable to deficiency."[17] Now, a Phillips redux, the New International Version: After three editions, his publisher wanted the fourth to attract more readers. So he asked Phillips to condense the book. Phillips made cuts, and then he wrote more. He also added his own design for a canal lock that would help drought-prone areas conserve water. Perhaps his editor found Phillips a challenge (the fourth edition came in at 600 pages). Phillips was enthusiastic, though, and enthusiasm is hard to resist.

On the cover, Phillips called himself "sometime surveyor to the canals of Russia under Mr. Cameron, architect to the late Empress Catherine II," and in the preface he gave more background: "My having been employed in England by the great Brindley, the father of English canals; having been a prisoner on parole for some years in America; and having travelled, with a particular view to observation on the state of their internal navigations, through Holland, Germany, Poland, to Russia, where I was employed by the government, will be allowed perhaps to furnish reasonable pretensions to a proper knowledge of the subject."[18] This intriguing resume would be worth a few detours. James Brindley alone deserves reconnaissance. But for zebra mussels, the Russian parts are closest to home—especially Section III titled, "Particular account of the canals in Russia; the first begun by Colonel Breckell, a German, who failed in the attempt, and fled the country in disguise; Captain Perry, an Englishman, was next employed by the Czar, Peter the

Great—Three different surveys of Captain Perry—Great extent of this Navigation, and amazing trade carried on it between China and Russia."[19]

Early on, Phillips made the key point that "all canals may be considered as roads of a certain kind."[20] Indeed. Our twentieth-century highways were his eighteenth-century waterways. Phillips was making a surface level comparison. On a surface of new liquid pavement, a fleet could carry more cargo: "In 1778 no less than 4927 vessels passed through the canal of Lake Ladoga, an increase of inland commerce of almost one-fourth in one year, by means of canal navigation."[21] A canal could connect distant places: "The communication, by water, between Atrascan and Petersburgh, or between the Caspian Sea and the Baltic, is formed by means of the celebrated canal of Vishnei-Voloshok."[22] A system of canals could expand an empire:

> Peter the Great, Czar of Muscovy, having observed, whilst in Holland, that the industrious inhabitants of that country had, by diligent perseverance, and principally by means of canals, raised a small tract of marshy land into a populous and powerful state; this great prince, among his other grand designs, formed the plan of having an inland navigation.[23]

As Phillips traced canals, he also traced the rise of the Russian Empire. In 1682, when Pyotr Alexeyevich became Tsar Peter I, the Swedish Empire controlled the Baltic Sea in the northwest, and the Ottoman and Persian Empires controlled the Black and Caspian Seas in the south. In 1721, when Tsar Peter I became Peter the Great, Emperor of All Russia, he had won from Sweden access to the Baltic Sea. Peter's Vyshnii Volochek waterway connected the Neva and Volga river basins, which in turn connected the Baltic and Caspian Seas. Within the waterway, the first Ladoga Canal bypassed Europe's largest lake (Ladoga), which was dangerous for shipping. The canal joined the Neva River, which entered the Baltic Sea at the Gulf of Finland. On the Gulf of Finland, Peter built Russia's new capital, St. Petersburg.

Canals and locks, dams and reservoirs, seaports: Peter's successors modernized and extended the empire's maritime reach. During the reign of Catherine II, Russia defeated the Ottoman Empire in the Russo-Turkish War of 1768–1774. The war was part of Russia's long quest for Black Sea access. Its concluding Kuchuk Kainarji Treaty stretched Russia to the Port of Azov and the northern Black Sea coast. Here Catherine's field marshal and former lover Prince Grigory Potemkin established the Black Sea fleet. From the Black Sea, Russia became a world naval power. Also from the Black Sea, it secured internal routes to ship grain from southern hinterlands. Russia grew again from three (forced) partitions that divvied the Polish-Lithuanian Commonwealth among its neighbors. Within Russia's split were the Oginski and Królewski (or Dnieper-Bug) canals.[24] Both linked the Baltic and Black Seas via the Dnieper. Meanwhile, Catherine's governor of Novgorod, Jakob Sievers, supervised reconstruction of the Vyshnii

Volochek waterway. Catherine appointed Sievers ambassador to the Polish-Lithuanian Commonwealth, to lead the second and third partitions. After her death, Emperor Paul made Sievers director of a new imperial Department of Water Communications.[25] And so on, as canals crisscrossed empire.

"When we consider the great labour, and prodigious sums of money those canals must have cost," Phillips marveled in 1803, "we shall be astonished that they were ever completed, as Russia has been almost continually engaged in expensive wars since they were first undertaken."[26] Phillips saw competition between canal building and war, but it was often the reverse: Canals were a feature of imperial expansion, with grand examples to come. From Black Sea to Baltic, Russia's waterways mapped the warring geography of empire.

All canals may be considered as roads of a certain kind. Phillips wasn't thinking below the surface. But below the surface canals were roads too. Below was an entire transportation system. Like the parallel universe above, below were main arteries (the Volga and Dnieper Rivers), destination nodes (Ponto-Caspian and Baltic seas), on ramps, off ramps, side streets, dead ends. Below came sprawl—the Mariinsk and Tikhvin waterways added to the Vyshnii Volochek, and later the monumental Volga–Baltic waterway replaced the Mariinsk. Zebra mussels were fit for this fast-changing geography. In an expansionary below, veligers could swim in new currents. Juveniles cemented byssal threads to ship bottoms. Traveling far from their homeland, zebra mussels were spotted in the Curonian Lagoon off the Baltic coast in 1804, on English docks in 1824, then Germany, Austria, Denmark, and the Netherlands.[27] *Dreissena polymorpha* became a species of immigrants.

Zebra mussels were positioned for a transatlantic crossing. Yet their first successful voyage was 1988, not 1888. The reasons why involve a more complex convergence of history, geology, and biology. The maritime histories of Europe *and* North America, the geologies of the Ponto-Caspian basin *and* the Great Lakes, the biology of dreissenids *and* unionids were all in play. The old world and new would have to collide. Once again.

In a 1973 landmark of environmental history, Alfred Crosby coined the term "Columbian Exchange."[28] Crosby argued that the year 1492 should not loom large in our historical psyche just because Christopher Columbus brought Spain to the Americas before Portugal, England, France, or the Dutch. No, the year 1492 should loom large because Columbus, his crew, his ships—the Niña, the Pinta, the Santa María—and the captains and crews and ships that followed unwittingly launched an earth-changing, history-making process: an exchange of plants, animals, and microbes between the old and new worlds. To sum up the argument, Columbus is important not only because he was a vicious conqueror in the age of empires but also because he was an early carrier of organisms in the Columbian Exchange.

Unlike Columbus, whose name endures, the twentieth-century captain who dumped ballast water full of veligers into the St. Clair River and Lake St. Clair remains anonymous. Still, we can imagine him and his ship as carriers in a later chapter of the Columbian Exchange.[29]

———

Who really killed the unionids in Lake St. Clair? Was it zebra mussels or humans? Geology or geopolitics? Nature or civilization? Over time such dualisms blur . . .

Stack layer upon layer of alternative stories, or self-serving viewpoints, to approach, perhaps, some truth: this is a narrative method embraced by filmmakers and historians alike. In his 1950 film *Rashomon*, Akira Kurosawa used the method so spectacularly that "Rashomon effect" became psychological shorthand for seeing partial truth but not *the* truth.[30] *Rashomon* was a mystery: Who murdered the Samurai? The solution was a shimmering mirage of memory, a wavy mirror that blurred subjectivity and deception.[31] Four witnesses, four realities, four waves across the sandy scene in the mirror. Rashomon referred to a historical place, for centuries the "magnificent and marvelous" southern gate to the ancient Japanese capital of Heian-kyo, now Kyoto, whose wooden "walls were white, and the pillars were vermillion, with genuine tiles on the roof."[32] By the medieval time of *Rashomon* the film, the gate was raw, ruined timbers, gape in the shell of a dying society, decayed shelter to drifters and demons and disposed dead bodies. Today nothing remains of the historical gate, not even a foundation stone, but in the film, Rashomon was a reef.

Don't hold fast to one point of view. Let go, *Rashomon* urges. Open your eyes to the scene. Then blink, and see it otherwise, then blink again. Who murdered the unionids?

———

What if zebra mussels were not despoilers who fouled native lands, or hardy wayfarers, or opportunistic migrants? What if they were just tiny animals navigating river currents? Blink, and see this reality: zebra mussels didn't think strategically, they *lived* strategically. Biologically speaking, zebra mussels are "r" strategists.[33] Remember the letter by thinking "r for reproduction"; as in, they reproduce rapidly, and in astounding numbers. One female can produce one million eggs every summer. The offspring of her offspring can number a half billion. Tribbles multiplying on the *Star Trek* Enterprise have nothing on zebra mussels. Like tribbles (those science fictional *r* strategists in outer space), zebra mussels eat huge quantities of food relative to their size—40 percent of their organic mass a day.[34] Except their food isn't a ship's store of grain but a water's store of "bioseston"— phytoplankton and other free-floating life. Absent predators, zebra mussels clean out the store, literally. They filter so much food that murky water filled with particles becomes clear; it becomes empty water, an aquatic food desert. In the Great Lakes, famine follows the zebra mussel. More on that shortly.

With life spans just four or five years, zebra mussels don't live long. But in their early years in the Great Lakes they lived hard, in superanimal bursts of power and motion. They blazed. Think six thousand candles burning on a square-foot cake. That's six thousand mussels on a square foot of rock, or a moving ship's hull. Six thousand mussels multiplied across space, multiplying through time, a living wildfire, radiating from Lake St. Clair up the St. Clair River to Lake Huron, blasting through the Straits of Mackinac to Lake Michigan, from Lake Michigan to the Chicago River and over to the Mississippi, then south to the Gulf Coast.

Merriam-Webster defines epicenter as "the part of the earth's surface that is directly above the place where an earthquake starts." A humble bulge in the river corridor physically separating Huron and Erie and politically separating the United States and Canada, Lake St. Clair became a kind of biocenter, the ground zero of a bioquake sending shockwaves through the Great Lakes.[35] Like famous earthquakes in human history, this bioquake produced an immediate "after" at the reef. As in, life as it was before the quake, and life as it was after the quake.

Feast and Famine

In a spider's web, what happens to the intricate mesh if something pulls hard on its anchor thread? Or the frame thread or a bridge thread?[1] Can we dissect the web to find out? Pin down a section of the mesh and tweeze its threads one by one? What about threads in a web of life? Each thread might teach us more about the web: ecosystem engineering, population dynamics, nutrient cycling, energy flows. We could spend years engrossed in a single thread before tweezing another, tracing it from here to there, or from before to after. But what if the examination is urgent?[2]

In the St. Clair waterway, zebra mussel colonies might have (hypothetically) benefited lake sturgeon. Dense mussel-mats created crannies for fish eggs, and new nooks where fry could hide. Adult sturgeon fed on mussels. Weaver and woven, mussels became bound in a web of life with indigenous fish like lake sturgeon and walleye, non-indigenous fish like the round goby, and miniature life like midges and mayfly larvae in the river and shrimp-like *Diporeia* spp. in the lakes (whose total tonnage could crush cities if dropped from space). When one tugged on a filament between them, others had to move. It would be sweet to introduce this wider community in the style of a children's show on marine life, a freshwater Bikini Bottom where eccentric people-creatures surround a lovable person-sponge. Sadly though, SpongeBob is the wrong visual. Think the remorseless food chain of King's Landing instead: a huge cast, labyrinthine relationships, reverberations after every event—a shape-shifter web of cause and consequence.

Below, follow a few spectral threads through the Great Lakes:

⸻ Zebra mussel→is food for the Ponto-Caspian round goby, which reached the St. Clair River in 1990→round gobies also eat lake sturgeon eggs and fry→fewer juvenile sturgeon survive at the reef.[3]

———Zebra mussel→outcompetes the native amphipod *Diporeia* spp. for settling algae (phytoplankton)→ *diporeia* populations crash in Lakes Michigan, Huron, Ontario, and eventually Erie→ lake whitefish depend on fatty nutritious *diporeia* but must shift to less nutritious zebra mussels→whitefish become malnourished and sickly→whitefish populations fall in Lakes Michigan and Huron, and sharply in Lake Ontario.[4]

———Zebra mussel→eats settling algae contaminated by mercury, polychlorinated biphenyls (PCBs), and other persistent pollutants→round gobies feed on zebra mussels→large fish consume round gobies→chemicals biomagnify up the food chain→top predators suffer reproductive problems→the round goby's new niche worsens the recycling of persistent pollutants.[5]

———Zebra mussel→filters the water column of phytoplankton and other particles (the water column is a deep-water pelagic zone, whereas the river or lake bottom is a benthic zone)→clear water with high visibility degrades habitat for pelagic fish like walleye, which need murky water→the Lake St. Clair walleye population falls→from plankton to top predators, the pelagic food chain collapses→macrophyte *Chara* becomes dominant at the bottom of the food chain, while visual predators like smallmouth bass and muskellunge become dominant toward the top→a benthic food chain replaces the pelagic food chain in Lake St. Clair.[6]

———Zebra mussel→filters the water column→sunlight penetrates deeper into the water→aquatic plants and benthic algae accelerate photosynthesis→in shallow Lake Erie plant biomass or "productivity" rises→zebra mussels excrete phosphate and ammonia, which fertilize the blue-green algae *Microcystis*, a cyanobacteria that produces toxins→algae eaters avoid *Microcystis* but overgraze more edible algae, leaving *Microcystis* dominant→the base of the food chain weakens→some summers, dense colonies of *Microcystis* bloom on Erie's surface→microbes break down a bloom's dead algae→decomposition depletes dissolved oxygen, making "dead zones" where organisms who can't swim to safety suffocate→scarcity and famine move up the food chain.[7]

———Zebra mussel→filters the water column and is food for round goby→Lake Huron alewives (non-native) lose phytoplankton food→Chinook salmon (non-native) put more pressure on alewives, their main food source→the salmon fishery and alewife populations crash→without alewives walleye rebound, and native lake trout now breed in the wild because alewives had produced an enzyme that hurt trout reproduction (by creating a thiamine deficiency in trout)→walleye, smallmouth bass, and lake trout prey on round goby→in Lake Huron, a new web with both native and non-native elements overlays an unsustainable salmon-alewife web.[8]

—— Zebra mussel→is food for round goby→in shallow water gobies swallow spores from the bacterium *Clostridium botulinum*→a neurotoxin in *C. botulinum* makes gobies sick with type E botulism→the fish get dizzy and swim erratically, they attract fish-eating birds like the common loon, which can dive two hundred feet for fish→poisoned birds become paralyzed, suffer respiratory failure, and drown→on shore, maggots eat the dead birds' flesh→gulls, piping plovers, and other shorebirds get botulism by feeding on infected maggots (fly larvae)→from 1999, 80,000 Great Lakes birds die→the round goby becomes a main vector for type E avian botulism.[9]

—— Zebra mussel→builds good habitat for other Ponto-Caspian life, not just the round goby but the quagga mussel, scud, spiny water flea, fishhook water flea, bloody red shrimp, and so on→they co-evolved in the ancient Ponto-Caspian seas, and adapt a Ponto-Caspian web to the young but compatible Great Lakes→frequent new arrivals reinforce the web→the Great Lakes undergo invasional meltdown.[10]

—— Zebra mussel→builds habitat for its larger cousin the quagga mussel→quagga mussels can live in deep cool waters and soft or hard surfaces→quaggas displace zebra mussels while expanding their range, and in Lake Michigan alone they reach 450 trillion mussels or 23 kilotons of biomass→quaggas filter the water column→in Lakes Michigan and Huron, quaggas trap phosphorus→they clean up phosphorus pollution from fertilizer runoff into Lake Michigan, but they also become a phosphorus sink→phosphorus recycling declines through the ecosystem, which means these quagga-infested lakes become "oligotrophic"→phosphorus drives ecosystem services, but phytoplankton and zooplankton no longer get enough→energy flows wane up the food chain[11]→the implication:

Less capacity to sustain life.

On Naming and Knowing

I fished the wall for the first time last year. I never caught anything but that realy didnt matter cause I was fishing. I did have one fellow come unglued on me. he said I took his spot. the thing is, that I was there a good 4 hours before him. I let him fluff his feathers and all and his woman tried to get him to calm down cause she noticed I had my kid with me. When he said F*** my kid I had to step up and extinguish his fire. His wife appologized for his behavior and he went and sat in the van. Other than that I met some good people and learned some valuable tecniques for fishing the wall. sorry for any miss spelling 😵

—Posted by Fishing 24/7 (his online handle) to a forum in
  Michigan-Sportsman.com.[1]

Fishing 24/7 was talking about a place in the Michigan city of Port Huron, where lapis-blue water meets ice-gray cement blocks at a seawall that lines and channelizes the St. Clair River. From the Blue Water Bridge (that crosses to Canada) to Pine Grove Park, two and a half miles south, runs a chain of skinny parking lots along the river. The lots' paint-platted parking spaces lie perpendicular to the river like French long-lot surveys of old, and make up a kind of accelerated waterfront property system. Each parked motorist has a temporary claim to river frontage with an unblocked view of the water and instant access to the seawall. For this reason, Pine Grove Park is a favorite for shore fishing off the wall, and in late spring, you'll see local anglers sitting on upside down white hardware-store buckets or Igloo or Coleman coolers with landing nets at hand, other belongings locked in the row of truck-homes behind them. Some of them chat while watching over heavy-action fishing rods clamped to the seawall railing and leaning over the fast current at the angle of wall-mounted flagpoles. Once in a while someone upsets the spatial diplomacy at the wall, as Fishing 24/7 called out in his post.

Forum members Puptent, Doubtndude, The Don and others gave Fishing 24/7 lots of righteous backup:

 There are a few who think they own a little piece of The Wall. Been fishing that part of the river most of my life, since the gravel pile days . . . but for the most part you are right, the people are pretty friendly! Plan on gettin the lines in Saturday!

Bottom line it is a public park and if your there first you get the best spots . . . there are great fishing spots along there, but with so many people out of work it tends to get crowded, but it is still PUBLIC. That goes for The Edison Parkway Boardwalk, Pine Grove Park Boardwalk, and the City Building/Sewage Plant Boardwalk. All open to fishing, all PUBLIC, and all first come. 😊

Bottom line make sure every one has room and hope they play well with others 😄

Flashy ripples of trucks and camo-capped anglers at the wall in Pine Grove Park offer a counterpoint to the grandiose but dormant real estate a short distance north before Lake Huron empties into the St. Clair. There, waterfront mansions and clustered "executive homes" loom like droid-houses over the moat of sand beach between them and the water. Weathered wooden docks look peaceable on the water's surface, but the sandy shoreline itself sports a hard armor of steel or rock groins installed in thin lines across the sand straight into the lake. Groin structures are supposed to secure and preserve each precious beach segment against attack from Lake Huron waves and changing water levels. A series of groins from property to property constitutes a "groin field." In a pamphlet titled "Living on the Coast: Protecting Investments in Shore Property on the Great Lakes," the US Army Corps of Engineers warns that armoring the shore with groins doesn't always work and may make beach erosion much worse for neighbors down-drift of moving sand.[2] So shoreline armoring can be an escalating and counterproductive affair. A few empty Adirondack chairs finish the deserted effect of people with buoyant employment, and boatloads of money, moored elsewhere till summer.

Back at Pine Grove Park, local anglers know that somewhere around 2.5 million walleye have finished their spawn in Lake Erie and are now running hard upriver, under the Blue Water Bridge, past all those vacant summer palaces, and into southern Lake Huron's deep water for a piscine summer sojourn.[3] After dark, or maybe during a windy cloud-covered day that makes for dirty water and great fishing (walleye don't like clear water), men and a few women fish off the wall hoping to intercept and lure any walleye leaving the current for a rest and a meal in the calmer water.[4]

In big fish lingo, the biggest walleyes caught earn admiring tags that range from man vs. nature à la the marlin in Hemingway's *Old Man and the Sea*— monster, giant, lunker—to man vs. man prove-it-to-me metrics—trophy walleye, double digit walleye (ten pounds or more)—to utilitarian farmer and his meaty livestock—pig, hog, jumbo.[5] The spring Marbleye Classic Walleye Tournament on the St. Clair River awards a five-hundred-dollar hog prize for "The Hog." The winning angler is that year's Hog Leader. Sometimes, though, St. Clair walleyes get the brotherly equivalent of shortening Roberts to Bobs or Phillips to Phils, in this case, walleyes to eyes, as in: "Plenty of eyes in the river right now. Guy next to us got a 31 inch eye off the wall. Buddy of mine bagged a dandy eye. Caught a couple of nice eyes. Fine looking eyes and that whitefish look great too."[6]

Okay, let's admit there are people who have no business calling a walleye "eye." Those of us outsiders uninitiated in the fierce intimacy of fishers and their fish, and who can't even pretend to relate to Santiago or Nick Adams, should probably refrain from sidling up to Doubtndude at the rail to casually ask, "caught any eyes?" We who find the siren call of iridescent creatures completely unhaunting, ungnawing, and resistible should turn back from the wind-ruffled river and go home before calling a walleye "eye." Any of us with the instinct to rescue family or friends from the Cult of the New Perfect Lure, and for whom lures named "chartreuse rubber" and "cotton candy paddle tail" sound incongruous with Cabela's, should never, ever call a walleye "eye."[7]

But there they are, the eyes in walleyes. Off the wall, walleye eyes add to the evening magic of this place for fishing. At dusk, anglers become tiny spectators to ghostly vessels, the famous one-thousand-foot lakers, each of which shades the river in moving blackness with reflections of stars and city lights reappearing in its wake until the colossal carrier glides over an indigo horizon. You can see across the St. Clair River to the city of Sarnia on the Canadian side, where countless amber lights glitter from Sarnia's "Chemical Valley," a dense concentration of petrochemical facilities—40 percent of Canada's entire chemical industry packed into fifteen square miles—that deceives the mind and spirit after dark by looking welcoming. In the water itself, just below the surface, glow uncanny luminescent orbs zigzagging in some quick pattern that eventually hypnotizes anyone who stares hard. These living flashlight beams shining up are eyes.

Stare at the word walleye, and let it stare back, not as a fish (for the moment), as a noun:

An eye with a whitish or bluish-white iris or . . . one with an opaque white cornea. Or, a condition in which the eye turns outward away from the nose.[8]

As an adjective, "walleyed":

Having eyes of an excessively light colour or showing divergence of some kind.[9]

There's more, there's an etymology, an ancestry of words, tracing *walleye* back eight centuries. In Old Norse, *vagleygr* is a compound of *vagl* and *eygr*, meaning "beam" and "eyed"—"beam-eyed." *Vagl* can also mean "film over the eye."[10] The later Middle English translation of the Old Norse *vagl-eygr* was *wawil-eghed*, just a few steps away from the Modern English word wall-eyed.[11] At some mist-covered moment, the English word became conjoined with a North American fish. Who mated word to fish? Who saw that the eyes of a walleye were wall-eyed, and that its eyes also glowed pearly?

It seems obvious that underwater fish don't need names to know each other. But above water, knowing begins with naming.

Walleye have gone by other names.[12] Francophone Canadians called the fish *doré* or "dory." Anglophone Canadians still call the fish pickerel, though this is confusing because there is an entirely different fish in the Great Lakes named pickerel. Sometimes Canadians and Americans bicker about this, with Canadians claiming the right to regional naming customs, and annoyed, confused Americans arguing for universal clarity about which fish they caught. There were also "yellow pike" and "blue pike," both abandoned in fish literature, and walleye was long called "walleyed-pike," even though it's not actually a pike, it's one of the perches.[13] However, the walleye's sleek, toothy silhouette is reminiscent of a pike, so it makes some sense that it was called "pike-perch," except for the separate European fish named pike-perch. The other pickerel, by the way, is a pike.

The walleye does have a universal name: *Sander vitreus*. This is its scientific name. No other animal in the world has this name. The Latin *vitreus* means *glassy*, *of glass*, *like glass*. "The eye of the living fish is like a glowing emerald," wrote ichthyologist (fish biologist) Tarleton Hoffmann Bean in his 1903 catalogue entry on walleye.[14] Again, back to the marble eyes of the Marbleye Classic Walleye Tournament.

Walleye have a membrane behind the retina of their eye that reflects light, like a mirror.[15] The reflection doubles the light that hits the photoreceptor rods of the fish's retina. This is how walleye can see in black water. Walleye eyeshine is white. When the fish is dead, its eyeshine glows like an icy White Walker's eyes, which makes one wonder if George R. R. Martin ever fished for walleye. One company designed a "Dead Eye Walleye" series of fishing rods. A cat's eyeshine is green; a rat's, red; a racoon's, yellow; a pine marten's, electric blue. Nature's own collection of iridescent glass marbles.

So even the scientific name *Sander vitreus* seems obvious in an "of course" way, as obvious and significant as when Gandalf the wizard realized the inscription "speak, friend, and enter" on the Doors of Durin told him the literal password for opening the doors; or, in twenty-first century terms, as obvious as the little yellow sticky note in the top corner of the computer monitor scrawled with phrases and numbers: "speak, child's first name and birth year, and enter." But like the caves of Khazad-dûm, there is no straight path if you venture into the branching caverns of name-giving through time.

———

Name-giving is the intellectual fishing grounds of the folklorist or historian of science. Such a historian could trace for us the origins of the name *Sander vitreus*, but might not know how to catch an actual *S. vitreus*.[16] Conversely, an angler fishing the St. Clair River might not know the factoid that, in 1818, "statesman-ichthyologist" Samuel Latham Mitchill presented *Perca vitrea* in the February issue of *American Monthly Magazine and Critical Review*.[17] (The name change from *Perca vitrea* to *Sander vitreus* comes later.) Mitchill's *American Monthly* submission of forty-two new species including *Perca vitrea* held the truly snooze-inducing title of:

> Art 1. Original Communications. Memoir on Ichthyology. The Fishes of New York described and arranged. In a supplement to the Memoir on the same subject, printed in the New York Literary and the Philosophical Transactions, Vol. 1. p 355–492. By Samuel L. Mitchill, [To the 166 species and varieties of fish mentioned in that paper, here is an addition of about forty more, making considerably above two hundred in the whole.][18]

But here is a flashier excerpt of Mitchill's *Perca vitrea* record in the magazine:

> GLASS-EYE—*Perca vitrea*, with the pupils of the eyes appearing like the semiglobes of glass in the decks of vessels, when illuminated on the opposite side, and with a yellow iris.[19]

Early nineteenth-century America was, among all the other transformations underway, a time of compulsive species searching and naming.[20] Globally this was an era of biological missions all over the world to find new animal and plant species. The first person to formally identify a species, like Mitchill with walleye, got to confirm it with a Latin name.[21] These scientific confirmations included records that described and classified each fish, and placed it in a never-finished Book of Life. By being first to publish a record of walleye, Mitchill linked his surname to the fish in perpetuity, so that any formal entry on walleye will read like this: *Sander vitreus* (Mitchill, 1818). You'll see the same format for the 205 other North American fish Mitchill recorded. Mitchill became renowned for this large number, but he wasn't alone. A fish rush was on.

The great collector and organizer of plant and animal names was Carl (Carolus) Linnaeus.[22] Linnaeus was history's uber name giver and "information architect," as David Quammen says.[23] In his 1735 *Systema Naturae*, Linnaeus wrote that "the first step in wisdom is to know the things themselves; this notion consists in having a true idea of the objects; objects are distinguished and known by classifying them methodically and giving them appropriate names."[24] Species naming and sorting—or taxonomy—wasn't new to Linnaeus or his time. In fact, Greek philosopher Aristotle was history's first taxonomist. But some 2000 years after Aristotle, Linnaeus "the great organizer" systematized what had become a chaotic process.[25]

Linnaeus's innovation was to limit any plant and later any animal to a two-part name in Latin, a binomial.[26] The first capitalized part of the binomial referred to an organism's genus, which was a biological subdivision with its own name giver and record of confirmation. The second lowercase part of the binomial applied only to the unique organism within that genus.

In the eighteenth century, Linnaeus would wait anxiously at home in Uppsala, Sweden for his apostles, as he called them, to send him new plant specimens from their biological missions to Asia, Africa, and the Americas. Because such travel was hazardous, many of Linnaeus's apostles suffered excruciating illnesses then died en route to these lands, or died once there, or died on their return.[27] Yet their calling to biological missionary work continued through naturalists and scientists in the nineteenth century, including Samuel Mitchill.

There is one telltale detail in the binomial *"Sander vitreus (Mitchill, 1818)"* that, sadly for those who want their fish history to be a simple journey there and back, leads to new caverns of rules and coded shorthand. Note the parentheses—"(Mitchill, 1818)." Such parentheses hold this specific meaning: that yes, Samuel Latham Mitchill gets credit for being the first to classify, name, and officially record the walleye *Sander vitreus*, but that *Sander vitreus* was not the original Latin name Mitchill gave. Mitchill's original name was *Perca vitrea*. If the original name had endured, there would be no parentheses around "Mitchill, 1818." Parentheses signal that at some point Walleye's scientific name changed. From *Perca vitrea* to *Stizostedion vitreum*, and finally to *Sander vitreus*, here were some of the stops along the way:

Genus: *Perca* Linnaeus, 1758: name-giver Carl Linnaeus
Family: Percidae Rafinesque, 1815: name-giver Constantine Samuel
    Rafinesque[28]
Genus: *Sander* Oken, 1817: name-giver Lorenz Oken
Species: *Perca vitrea* Mitchill, 1818: name-giver Samuel Latham Mitchill
Genus: *Stizostedion* Rafinesque, 1820: name-giver Constantine Samuel
    Rafinesque
Species: *Lucioperca americana* Cuvier, 1828: name-giver Georges Cuvier
    (name later rejected)
Species: *Stizostedion vitreum* (Mitchill, 1818): walleye moved to the
    Stizostedion genus
Species: *Sander vitreus* (Mitchill, 1818): walleye moved to the *Sander* genus

To understand how dynamic this naming business was, look at Mitchill's *Perca vitrea* and Cuvier's *Lucioperca americana*.[29] Mitchill and Cuvier gave the same fish different names. Eventually it was sorted out in favor of Mitchill because his 1818 classification came before Cuvier's 1828 classification. But there were decades when both names were in use.

Cuvier's *Lucioperca americana* remains immortalized in one of the first significant publications on the Great Lakes, Louis Agassiz's 1850 *Lake Superior: Its Physical Character, Vegetation, and Animals, Compared with Those of Other and Similar Regions; With a Narrative of the Tour, by J. Elliot Cabot*.[30] *Lake Superior* was a combination scientific publication (Agassiz) and journey travelogue (Cabot) of their 1848 expedition from Boston to Lake Superior.

At the time of the expedition, Agassiz was a Harvard professor of geology and zoology and renowned in both realms. Now called the "father of glaciology," he was the first to theorize a past Ice Age on Earth during which ancient ice sheets advanced across continents and sculpted modern landscapes. Agassiz was also a former student of Cuvier, and a foremost ichthyologist of the day. Cabot's first-hand portraits of him show an intense drive to know the natural world up close, to "know the things themselves," as Linnaeus said.[31] Agassiz biographer Christoph Irmscher concurs: Agassiz was "*doing* science wherever he went," he says, "his pockets . . . stuffed with specimens."[32]

To reach Superior, the company journeyed up the Great Lakes to Sault Ste. Marie.[33] The Huron–Erie corridor of our spring walleye run was one leg of that journey. Today's anglers fishing off the wall of the St. Clair River might like reading about this part of the trip because Cabot wrote scenes in such photo-realistic detail, and also because he was funny. Cabot seems to have been underwhelmed with the so-called "greatness" of Lakes Ontario and Erie, which did not live up to their "glorification" and were, he complained, "tame as the edge of a duck pond."[34]

As the river shallowed into a delta entering Lake St. Clair, they encountered a more extraordinary waterscape, the St. Clair Flats, the largest freshwater delta in North America, which was, as Cabot described it, a "remarkable extent of mud-flats, (some twenty miles across,) covered with only a foot or two of water in most parts, and even the channel is so shallow that the larger boats have to discharge a part of their cargo into lighters while passing it."[35]

Vivid as Cabot was, he didn't do justice to the St. Clair Flats. The flats of their voyage deserved what Anne-Marie Oomen calls "an act of ecocomposition."[36] For that we can look to Constance Fenimore Woolson, grandniece of James Fenimore Cooper. As a teenager Woolson had traveled the Great Lakes with her family. Years later she drew on the lakes for short stories, including "The St. Clair Flats." Woolson set the story in 1855, about the time she saw the St. Clair Flats, and just a few years after Agassiz and Cabot. Her unnamed narrator, on a fishing excursion from Detroit, tries to compose the delta in words as the steamer slips into its shallows:

> The word "marsh" does not bring up a beautiful picture to the mind, and yet the reality was as beautiful as anything I have ever seen,—an enchanted land, whose memory haunts me as an idea unwritten, a melody unsung, a picture unpainted, haunts the artist, and will not go away.[37]

Later, the narrator's companion Raymond extols:

> What a wild place it is! . . . How boundless it looks! One hill in the distance, one dark line of forest, even one tree, would break its charm. I have seen the ocean, I have seen the prairies, I have seen the great desert, but this is like a mixture of the three. It is an ocean full of land,—a prairie full of water,—a desert full of verdure.[38]

And what of the physical complexity of the delta? Here again Woolson provides the visual:

> On each side and in front, as far as the eye could reach, stretched the low green land which was yet no land, intersected by hundreds of channels, narrow and broad, whose waters were green as their shores. In and out, now running into each other for a moment, now setting off each for himself again, these many channels flowed along with a rippling current.[39]

Until the first deep dredge of the St. Clair delta, making an industrial canal for ship traffic, Woolson's desert full of verdure was Cabot's remarkable extent of mudflats.

The Agassiz boat would continue past St. Clair riverbanks that today are a cement seawall. Missing from the 1848 riverbank were Ojibwe villages. In 1807, the Treaty of Detroit had ceded Indian lands in southeast Michigan to the United States. The treaty established two reservations in St. Clair County for local Black River and Swan Creek Bands of Ojibwa, but nearly thirty years later the 1836 Treaty of Washington broke up the reservations and forced the two bands out of the state or farther north to a combined reservation with the Saginaw Chippewa.

In 1998 testimony supporting federal legislation to recognize the sovereignty of the Swan Creek Black River Confederated Ojibwa Tribes of Michigan, historian Deborah Davis Jackson offered her stark assessment of these forced removals: "As the nineteenth century progressed, the Swan Creek Black River Ojibwa people of Southeast Michigan found their options drastically restricted by the actions of the Federal Government (confining them to reservations, then extinguishing those reservations, then physically lumping them together with the Saginaw Chippewas on a reservation far from their homeland)."[40] Today the State of Michigan recognizes the tribe, but so far neither the federal government nor the Saginaw Chippewa do.

Back to 1848 and a rainy June 23 in Lake Huron at "Mackinaw," or Mackinac Island. Here the Agassiz group missed the last steamer to Sault Ste. Marie and had to hunker down at the Mission House Hotel. The Mission House was a former Presbyterian boarding school for Indian and Métis children in the upper Great Lakes.[41] The island's Mission Point was still a gathering place for Ojibwe and Odawa, but, said Cabot, "we were disappointed at finding only three or four lodges of Indians here. In August and September (the time for distributing the "presents,") there are generally several hundreds of them on the island."[42] Cabot's offhand and touristy comments are jarring and offensive now, when we know that for hundreds of years Indigenous peoples occupied Mackinac Island and gathered to fish and to catch up with their own extended families and geopolitics, and when we also know that the Mackinac Island Odawa chief and fur trader Agatha Biddle was one of the most influential people in the region before and after the Agassiz visit.[43]

For his part, Agassiz focused on fish. "Notwithstanding the rain," Cabot gushed, "the professor [Agassiz], intent on his favorite science [ichthyology], occupied the morning with a fishing excursion, in which he was accompanied by several of the party, most of them protected by water-proof garments, while he, regardless of wet and cold, sat soaking in the canoe, enraptured by the variety of the scaly tribe, described and undescribed, hauled in by their efforts."[44]

Cabot continued: "With a view of indoctrinating those of us who were altogether new to ichthyology with some general views on the subject, he commenced in the afternoon, scalpel in hand, and a board well covered with fishes little and big before him, a discussion of their classification."[45]

Then Agassiz delivered a mini field course on walleye taxonomy based on Cuvier's classification, *Lucioperca americana*. Agassiz began, "The fish before us belongs to the genus *Lucioperca*. They have a wide mouth and large conical teeth, like the pickerels, and two dorsals. There are two species in Europe and two in the United States. This is *L. americana*; its color is a greenish brown above, with a whitish below, and golden stripes on the sides."[46]

Agassiz overlooked the eyes of his walleye. Had he examined *"Perca vitrea"* instead of *"Lucioperca americana,"* would the presentation have changed? Mitchill's name and published description of *P. vitrea* made the fish's stunning eyes a focal point, while Cuvier's name and published description of *L. americana* said nothing about eyes.[47]

Even apart from the Cuvier-Agassiz walleye *Lucioperca americana* and the Mitchill walleye *Perca vitrea*, what about the specie's nomadic journey from the genus of *Perca* to *Stizostedion* to its final name-home in *Sander*?[48] *Perca vitrea* might not have endured as a name because, sometimes, a genus itself does not withstand scientific tests of time. Or, alternatively, a species might move from one genus to a different genus as the classifications improved. This was the case for walleye.

Today geneticists carry out genetic stress tests to achieve still more precise models of subspecies, species, genera, subfamilies, families, suborders, orders, classes, superclasses, subphylum, and phylum. Genetic testing prompted the walleye's last move to the genus *Sander* by proving its close relationship to a European fish already in *Sander*.[49] Since *Sander* was senior to *Stizostedion* by three years, *Sander* became the genus.

A quick digression for fish foodies: Beware!—the European *Sander lucioperca* (Linnaeus, 1758) is such a near relative to walleye that Minnesota and Wisconsin restaurants once substituted it for walleye without telling customers. This was a bold cultural sacrilege when the official state fish of Minnesota is the walleye and five Minnesota cities proclaim themselves the Walleye Capital of the World.[50]

Here is walleye's classification now:

Kingdom
Animalia
animals

Phylum
Chordata
chordates

Subphylum
Vertebrata
vertebrates

Superclass
Gnathostomata
jawed vertebrates

Class
Arctinopterygii
ray-finned fishes

Order
Perciformes

Family
Percidae
perches

Genus
Sander

Species
Sander vitreus
walleye

———

Sander vitreus epitomizes the scientific field of taxonomy.[51] From Linnaeus to now, taxonomy became the sum of a perpetual information enterprise to describe, name, and classify the living world into a hierarchy of relationships, to "create cosmos out of chaos."[52] The whole of taxonomy has been a collective, centuries-long biopedia, a proto-Wikipedia, mass production of an ever-expanding Encyclopedia of Life. Marine ecologist Boris Worm calls it "Nature's Library."[53] But 250 years after Linnaeus, the library's holdings account for just 15 percent of life on Earth. "We've only begun to decipher the first ten books," says Worm.[54]

Think about it; we don't know 85 percent of all life. Two hundred fifty years of knowledge generation and we're still ecologically illiterate. Add grim context, that we're in the midst of a sixth mass extinction, this one human caused, and Worm warns that, "we're throwing out entire books without having a look at them."[55] Legendary biologist E. O. Wilson is even more urgent about species and biodiversity collapses. Wilson calls for a renewed and accelerated taxonomy to confront the crisis.[56] In "Engineering a Linnaean Ark of Knowledge for a Deluge of Species," taxonomist and entomologist Quentin Wheeler proposes a global coordinated cyber-taxonomy, "a taxonomic moon shot."[57] He envisions "a societal commitment to intensively invest in species exploration for the next fifty years. Why not aim for 200,000 species per year within two years, ramping up to 500,000 species 'treatments' per year within a decade?"[58]

Yes, why not? May hundreds of obsessive taxonomists enlarge the known population of Noah's Ark. In the words of Louis Agassiz's close friend Henry Wadsworth Longfellow in "The Building of the Ship":[59]

Build me straight, O worthy Master!
Stanch and strong, a goodly vessel,
That shall laugh at all disaster,
And with wave and whirlwind wrestle!

It's easy to cheer and be cheered by such belligerent optimism. Pugnacious, never-give-up optimism may be the best way to face waves and whirlwinds—species extinctions, water wars, climate change.

Yet global-sized mudflats and sandbars impede the route that Wilson and Wheeler chart. In the eighteenth and nineteenth centuries, the "Linnaean enterprise" of species seeking attracted intense public interest. But no encyclopedia has inspired a worldwide quest to actually protect life on Earth. Will 500,000 species "treatments" (meaning 500,000 classifications of newly discovered life) arouse public passion in the twenty-first century? And not to be too harsh, but "treatments" is a cold word. "Treatments" does not persuade the mind or the spirit.

A modest suggestion for a scientific movement that aspires to help save life on Earth: Reclaim the art and poetry of science. In the early days of taxonomy, scientists had professional and spiritual partners. For one, the science-art symbiosis was powerful. To the ceiling, Linnaeus plastered the walls of his own study with botanical prints from his collaborator, the illustrator Georg Dionysius Ehret.[60] Linnaeus literally surrounded himself with the art of taxonomic classification.

The visual arts were intrinsic to the science.[61] To train as a naturalist meant practice and proficiency in illustration. No expedition set off without at least one illustrator, and classifications included precise drawings of plant or animal. Sometimes these were gorgeous. Charles Alexandre Lesueur, another of Georges Cuvier's students and the first naturalist to study Great Lakes fishes, was a celebrated artist: "His descriptions are clear, exact, and honest. His drawings are not accurate only, but spirited. They are works of art rather than mechanical representations."[62] Meanwhile a rapt public embraced wondrous illustrations of the natural world—delicate lines, hand-tinted shadings—each a coming-out celebration for a spectacular life form in all its strange, complex beauty.[63]

Farther afield than illustration, scientist and poet could be symbiotic, and so could scientist and philosopher. There was no "great divide between science and humanities," as historian Jessica Riskin notes.[64] Reason engaged imagination; theory engaged creativity so that, in the eighteenth century, Johann Wolfgang von Goethe—Germany's great poet, playwright, novelist—dared critique Linnaeus's classification system as mechanical and static, and published his own minor scientific treatise, *Metamorphosis of Plants*.[65] For Goethe, science was art, and the experimenter was transformed by the experiment. In the nineteenth century, Agassiz's expedition colleague James Elliot Cabot was a philosopher. Agassiz and the poet Longfellow nurtured their blended intellects and identities. "No one loved Agassiz more than the poets and philosophers," according to his biographer. "He was Emerson's transparent eyeball personified: the divinely entitled eye."[66] As in Ralph Waldo Emerson's *Nature*, a landmark of nineteenth-century American transcendentalist philosophy:

> In the woods, is perpetual youth. . . . Standing on the bare ground,—my head bathed by the blithe air, and uplifted into infinite space,—all mean egotism vanishes. I become a transparent eye-ball. I am nothing. I see all. The currents of the Universal Being circulate through me. I am part or particle of God.[67]

Such synergy crackled and sparked, and jolted a nineteenth-century public into seeing nature and the earth in profound new ways.

Polymathic exchanges and friendships continue in the present. Poet Aimee Nezhukumatathil draws on the "musical language and diction of science."[68] Poet and philosopher David Whyte began his career as a marine zoologist. Scientific illustrator Joseph Tomelleri astounds with his fish drawings. Digital 3-D models take animal sculpture in fantastical directions. Look up "electron microscope photography," and behold. My own colleague, ichthyologist Devin Bloom, sees the fusion as more intentional than ever: "I know many scientists who are artists and paint, draw, sculpt, and otherwise enshrine the natural world and the organisms they love," he says. "Many peer-reviewed publications have special covers with gorgeous art and sections at the end of papers and issues with more art of varying sorts. We take photographers, videographers, abstract artists, and more on research expeditions to help us capture the beauty of the natural world. I have friends who are scientist-musicians and write music about nature.[69]

Yet Bloom agrees that our larger cultural vision has dimmed for seeing these permeable swirls of knowing and insight among scientists, artists, philosophers, and poets. As a case in point, most college students in the sciences won't experience a sustained Agassiz–Emerson synergy *during* their four or five years at a university. They won't find art embedded in or intrinsic to their majors. That's not how their learning is structured. How many biological science departments require applied art training as part of the curriculum? Likewise, humanities or art students are unlikely to see themselves as contributors to new knowledge about the natural world, much less an integral part of any scientific field.

Four years is a long time to experience compartmentalization firsthand, and more than enough time to internalize methods of compartmentalization. Fair or not, here's a blunt-force historical reading that may reflect a student's experience as a proto-scientist or proto-artist or proto-poet: From the nineteenth to twentieth centuries, formal disciplines within the natural sciences took shape, including botany, zoology, ornithology, ichthyology, entomology, and so on. These hardened into detached knowledge-generating structures, with walls to define and confine and promote the work within. Art, poetry, and philosophy were left outside the walls.

Devin Bloom points to another reality. "During the heyday of exploration, scientists were members of high society and the public engaged them. They were invited to the fancy parties and given a forum. Scientists don't get to decide if the natural world takes a central role in life, art, and spirituality, everyone else does."

Agreed, and so the swirls go round. More than ever, natural scientists need partners outside their labs and departments—or just outside—kindred spirits, alter egos, who know the natural world differently but consider themselves equals in the exchange.[70] More than ever, artists, philosophers, poets should share rooms with science comrades, both walled workaday rooms and Earth's unwalled rooms on water or land.[71] And certainly, taxonomists need strong backup in translating the existential meaning and transcendence of 500,000 "treatments."

———

One last thought on the Linnaean route to conservation: By focusing on name-giving, on taxonomic rules, on mass information production, does a walleye blur into abstraction, and become veiled from our mind's sight? Do people fade, too? Within the annotated and italicized lists of knowledge, there is no seawall in Port Huron, no anglers tussling over fishing spots, no eerie eyes glowing below the water's surface on a cold spring night. In walleyean nomenclature, the physical world was central to naming and knowing: *physically* known by the fisher who tugged a powerful but frantic body at the end of a hook or bottom of a net; *directly* known by the scientist who turned over a corpse, a prize specimen to admire and draw, to dissect and examine, and eventually to preserve in a specimen collection among other corpses.

And perhaps, someone ran their hand over the living fish as they said its name, a caress from head to tail along the slick nap of interlacing scales, to calm the alarmed spikes of the dorsal fin, and maybe, for a split second, to calm the spirit of a fellow traveler running hard upstream.

Part II. On Seeing and Knowing

AN UNDERWATER BIOGRAPHY

Bad Diver

Should you ever find yourself in the Michigan city of Port Huron, take a walk away from Lake Huron, and go south along the seawall of the St. Clair River. The seawall lines and constricts the St. Clair until it finally exhales miles downstream into the open water of Lake St. Clair.

Anglers at the wall might nod at you, then stand aloof. Looking out at the river you'll probably note the silver-blue glitter of water dancing on the river's surface. And the skyline of smokestacks a hundred yards across the river, in Sarnia, Ontario's Chemical Valley, a rust-red freighter anchored in its harbor. But looking into the water, here's what you and I would not see: We would not see the shifting silts and gravels of the river bed as they expose a slim corked bottle. Rolled up inside is a hundred-year-old steamship deposit slip (White Star Line #32616, June 30, 1915), along with a perky message from Tillie and Selina, young women on the steamer. "Having a good time at Tashmoo," they call across the century.[1] Tashmoo Park was an early amusement park on the St. Clair Flats, at Algonac.

We wouldn't see the scuba diver creeping across the bottom of the St. Clair River to the US side, eight pounds of marijuana stuffed in a green tube belted to his wetsuit.

Or the World War I bomb.

The Enbridge oil pipelines.

Barbie dolls. Dildos. Dioxins.

Ceramic shards, sunglasses, cell phones, shipwrecks.

Or this:

"Detroit Archdiocese Says Missing Church Found under Lake St. Clair."[2] Yes, a church once went missing in Michigan's most populated region.

It seems that underwater realms are human realms, too. The water reflects *us* back to *us*.

As for that scuba diver . . .

There was more action below than international pot-running. Where water approached the wall at Pine Grove Park, shore anglers got aggravated by all their lost fishing lures. At times the site was a Bermuda Triangle of walleye tackle. Here's what the anglers couldn't see thirty feet beneath the surface: aluminum window screens on the river bottom, ripped and gouged on purpose, so the screen would snag their barbed lures. A masked man (dressed head to fin in black) had positioned these scragged screens where he could maximize his catch. He would return for the lures another day.

Greg Lashbrook called this person a "bad diver."

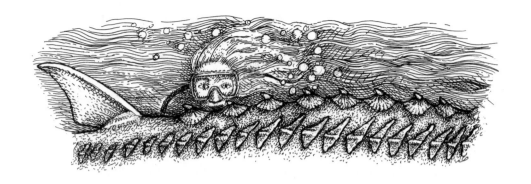

Waters That Bind

Greg Lashbrook and Kathy Johnson make up the Port Huron diving and underwater film team "Gregory A. D.," or Gregory Art and Diving.[1] In his sixties yet ageless, Greg looks tough, with a sturdy build, grizzled beard, sprawling sideburns, brown-brillo hair escaping to his neck from a faded canvas tie hat, which is a kind of skull cap tied tight in the back. Greg shows me a prosthetic hook, like Captain Hook's hook. He collected this from the river bottom and added it to his heaps of St. Clair treasures—found objects for some future artistic 3-D creation fusing Greg's fantastical inner vision with the material world of the river. My favorite is the mask he molded from the lead of two-pound fishing sinkers. These triangular hunks of ecological kryptonite are a renewable resource in the river. Melted then mutated, the lead forms a mottled skin of silvery facets, pockmarked in places with inlaid pennies and nickels. A river god's caustic carnival mask. Orange mirrors in Oakley sunglasses hide the eye slits. Rusty iron coils shoot fifteen inches high from the mask-head like kinky strands of hair. The coils were not originally hair, Greg divulges, they were the brain erupting in orgasm.[2]

The front of Greg's cap has an embroidered skull-and-crossbones. Yes, exactly, Greg looks like a pirate.

Kathy and Greg begin the story of their 1990 wedding for still another incredulous land-lubber. *To be that completely at home underwater? . . .*

When they tell a story together, you have to beat back the cliché "opposites attract." Twelve years younger than Greg, Kathy is lithe, even fragile, no makeup on delicate skin, wisps of light brown hair that streak blonde in the sun. Kathy is also vivacious, voluble, responsive in a nanosecond. Greg is cautious at first. He lets the air pressure drop before a story rolls forth. Their voices hit different keys: Kathy's is viola-smooth, although she punctures the music of her voice with snare-drum laughter. Greg's is the cello. The musical term "counterpoint" is probably a better cliché for them than opposites. There's an edge, though. Kathy manages their business, while Greg is the artist. "A lot of people call me a bitch," she says, "but everybody loves Greg." Figures, I think on Kathy's behalf.

They held their wedding at the Lake Huron shipwreck *Sport* (b. 1873, d. 1920). The *Sport* was the Great Lakes' first steel tugboat. Today it rests on Huron's lake bottom, forty-five feet below. A wetsuit manufacturer custom designed Kathy's white neoprene wedding dress and Greg's black-and-white suit, complete with bow-tie and tuxedo tails. Kathy wore a veil.

They first met in an indoor swimming pool. Greg was Kathy's assistant dive instructor in a scuba class. This was 1982, Kathy was sixteen. "Oh my!" she thought when she saw him. Greg "did this beautiful dive," she reminisces. "He would land more than half-way across the pool before he even hit the water, and then he would use that momentum to carry himself right to the wall. He didn't even really swim. The whole thing was very fluid. He never stopped moving." Later Kathy returned to classes as an assistant herself. She continued to admire Greg's grace and intuition. She gives the example of one of the most important days of an introductory scuba class, when students sit on the bottom of the deep end of the pool to practice flooding their masks.

"Every time, one or two of 'em freak out, cause it's a really weird sensation," Kathy says. "You lose your sight. Losing sight when you're under water, your brain is like, *that's it, you know I'm done, I already thought this was a dumb idea to be underwater trying to breathe air, this is not natural, get me outta here!!* So panic sets in, and they pop. We're there to slow down their pop and make sure they don't hurt themselves. And it was so uncanny. He knew. He was next to them on the bottom before they popped, every time."[3]

Greg nods. "It's all in the eyes," he says, "all in the eyes."

Two years after Kathy's first scuba class they began dating. Theirs was not a wait-until-she's-of-age love story. Their romance began because Kathy's family had a house on the St. Clair River. "He really did scope classes for *anybody* who lived on the river, like if *aaanybody* came through that scuba class who lived on the river he was their new best friend."

Eyeing her playfully, "Well, it helped if you were a girl."

Kathy continues that Greg was always on the lookout for new spots to dive from shore, because most of the riverbank was private—lots of expensive homes—and most owners didn't allow access and, she explains, "a lot of those places were 'virgin,' meaning they had bottles laying on the bottom since the 1920s or even the turn of the century."

"So then he started coming to our house to dive."

The circuit judge who married them was a diver. Diving friends and colleagues were father of the bride, best man, wedding party. They used little microphones in their face masks to take their vows. The local newspaper couldn't resist a stream of puns—"wet wedding," "Neptunal nuptials," "couple to take plunge to tie the knot," "the bride who couldn't care less if it rains on her wedding."[4]

Greg and Kathy cobbled together a living. Greg was a union journeyman, so he could work construction as a carpenter or machinist. But "I wasn't that good of a carpenter," he admits, "I'm a much better artist. I'm pretty much afraid of heights once I get over thirty feet. I can't walk a beam. I tried, I practiced on my lunch hours, and could not do it."

"It's embarrassing to have to slide across on your butt," he adds.

Early on, they concentrated on commercial diving (inspecting underwater infrastructures), rescue diving (for the marine patrol of the St. Clair County sheriff's office), stock photography (nature and marine images).[5] Greg was a versatile artist, so they took commissions from local businesses, or they bartered—a painting, a sculpture, an etched glass window design in return for food, a household item, a service. They received critical dental work through barter. For a long time, most of their income came from T-shirts they screen printed with Greg's designs. They sold the T-shirts for ten dollars. The county sheriff framed one that Greg made specially for him.

They had setbacks. Kathy got Crohn's disease, and they were dropped from their health insurance plan.

One year Greg answered a small ad in the back of *Sail Magazine* (or was it *Cruising World*?). The job entailed two winters crewing, from Galveston, Texas, across the Gulf of Mexico to the Yucatan Peninsula, "in a boat that wasn't properly designed or built for ocean travel," Greg learned in transit. The two of them considered relocating to the tropics, maybe the Caribbean, where they could focus on diving, and photographing or filming marine life. "We reached a level of comfort under the water, so that when we add another task like filming [Greg] or modeling [Kathy], we can do that, we can work." They had expedition experience. They felt skilled, resourceful, frugal enough to follow that path. But ultimately, they turned away from the attractions of sun and saltwater, and gorgeous tropical fish. Greg wanted a home base near Port Huron.

Kathy pauses, and then, "he *has* had a longstanding affair . . ."

Huh?

"He does have a mistress, and it's the St. Clair River. She lures him back no matter how far away he goes."

Greg smiles, then muses on that. "There's a lot of good diving in this world. But there's no place I've found like the St. Clair River. The combination of current and clear-water and marine life and freighters over your head combines to create a really exciting experience."

"Even our friend from National Geographic said that he's never dove anyplace else like it," Kathy finishes.

In 1991, they published a book for the international series of Pisces Diving and Snorkeling Guides: *Diving and Snorkeling Guide to the Great Lakes: Lake Superior,*

Michigan, Huron, Erie and Ontario.[6] But their real turning point was an environmental controversy on the Black River, a large tributary of the St. Clair River. Intense flooding on the Black had created an outcry from Sanilac County farmers, who wanted the US Army Corps of Engineers to carry out a flood control project. This included dredging 24.3 miles of the river. The river was also home to an endangered unionid, the northern riffleshell clam (*Epioblasma torulosa rangiana*). The northern riffleshell occupied a miniscule five percent of its historic range. In Michigan, their only populations were in the Black, St. Clair, and Detroit Rivers.

Sanilac County agreed to pay for divers to relocate the clams from the Black River to the Detroit River until the dredging was complete, and then to return the clams. The Port Huron-based Commercial Diving and Marine Services Inc. won the bid to relocate the clams. Company owner Wayne Brusate was one of Greg's early co-workers and close dive buddies, and also dive chief for the sheriff's dive unit.[7] Brusate's idea for the clams was to use gold dredges to suck in debris, but instead of sifting for gold, the company's dive teams would sift for clams. In summer of 1988, Greg and Kathy were one of the two teams. Their job was to bag every clam, hold them in the river for biologists to inspect, then install them in the Detroit River. "We got where we knew every species of clam," Kathy recounts.

"The environment and these different species," Greg weighs in. "I always wanted to go to work for Greenpeace and be the one on the boat in front of the harpoons. I always thought I wanted to do that. Not that this is anything like that, but we seemed to have kind of made a niche doing that sort of stuff." Only their niche is in the Great Lakes.

The northern riffleshell relocation did not work. All 118 clams died. A Michigan State University Extension publication stated as fact that the Black River transplants "survived for three years until zebra mussels invaded the area and colonized the unionids."[8]

Well, not exactly.

Zebra mussels were not why those *specific* clams died. Scientists had chosen a relocation site in the Detroit River close to a remnant population of northern riffleshells.[9] Greg and Kathy built a little corral with chicken wire on the river bottom, thinking it might help the company identify the Black River clams quickly. (Today they laugh at their younger selves, because corralling clams was completely unnecessary.) Two years later, when Commercial Diving and Marine Services went back for the clams, to return them to the Black River, they found that "someone had dredged upstream in the Detroit River and sedimented out the whole site, and killed them off." In Kathy's wry words, "a bad footnote to that research project." "It wouldn't have mattered," Greg points out, "because the zebras finally came in and devastated everything."

In the Great Lakes, conservation is a rough business. Everyone in the business knows this. Yet even failed projects can be guides. After the northern riffleshell job, the duo chose not to pursue the most common opportunities for Great Lakes divers—diving instruction and outfitting, commercial diving for industry, and especially not shipwrecks. "When people ask us about wrecks we try to be clear: Nope, don't know anything about wrecks, don't really care even. But you wanna talk about fish? That's our passion."

Gregory A. D. carved out an environmental niche as "freshwater marine life specialists." For the D in Gregory A. D., they focused on conservation diving and freshwater filmmaking. The conservation part connected them to university and agency researchers and community organizations. For a long time they liked working with biologists "who were interested in what was happening underwater in the Great Lakes." Eventually they became cynical about scientists and freshwater science publishing, but that's a later story.

Freshwater filmmaking was more precarious at first. In the 1990s, "there was no interest. We couldn't sell freshwater footage to the Discovery Channel to save our life." By the 2000s, fascination with the Great Lakes had grown, and Gregory A. D. did work for the Discovery Channel, Animal Planet, National Geographic, the IMAX film *Mysteries of the Great Lakes*, even for a show about aliens. "There's apparently something weird kooky hooky going on over the middle of Lake Michigan," Kathy says. For the kookiness, look up "Lake Michigan Triangle."

Kathy and Greg's most fulfilling project was filming and producing the 2011 documentary, *Manistee Nmé: A Lake Sturgeon Success Story.*[10] Nmé is the Anishinaabek name for lake sturgeon. In the film, they chronicled the pioneering work of the Gaaching Ziibi Daawaa Anishinaabe, or Little River Band of Ottawa Indians, to restore a nmé population on the Big Manistee River, which empties into Lake Michigan. Said natural resources director Jimmie Mitchell of the project, "The Tribe, through its sturgeon-rearing program, is striving to make amends for 170 years of destruction done against this noble species."[11] Above water, Greg and Kathy followed the tribal community and tribal biologists. Underwater, they followed nmé.

The Gaaching Ziibi Daawaa Anishinaabe (herein LRBOI, following the band's fisheries publications) had a cultural kinship to nmé that shaped the science of the tribe's restoration program.[12] The LRBOI prohibited many common fishery approaches of that time (2001). These included methods they thought would cause pain or distress to the fish: No setlines with hooks to capture sturgeon. Or methods that would violate a sturgeon's autonomy; slicing its belly to insert a transmitter for tracking was out of the question. Or methods that would interfere with the population's own natural selection during spawning. Biologists had to figure out a way to retrieve naturally fertilized eggs from the river. This last was one of the most definitive breaks from standard fish hatchery procedures.

The standard approach had been to physically extract eggs from a female, then select a handful of males for their sperm. If sperm from four or so males fertilized the female's eggs, hatchery-raised sturgeon would retain genetic diversity. But two implications caused the tribe to do things differently: First, in the standard approach, fisheries biologists would make the reproductive choices, as opposed to the sturgeon. And second, egg extraction would involve brute force over the female egg donor. As Greg and Kathy once observed in a Wisconsin river, a team would pull the female out of the water. Laying the sturgeon on her back, three or more strong people would pin her writhing body. Another person would press their full weight on her belly, pushing hard to force the eggs out.[13] Someone else would be ready with a stainless-steel bowl to collect the eggs.

The LRBOI rejected these procedures. Instead, its biologists devised less invasive methods for working with individual fish and with the larger nmé population. Natural fertilization in the river was a must.[14] "Lady's choice," as Kathy puts it. This wasn't a new idea among Indigenous peoples. Louise Erdrich writes about early Ojibwe sturgeon gardens at Lake of the Woods, in Minnesota—"shallow and protected parts of the lake where they mixed eggs and sperm and protected the baby sturgeon from predators. The eggs and sperm were mixed together with an eagle feather in an act both sacred and ordinary."[15]

One LRBOI innovation was a simple, cheap way to acquire fertilized eggs from the river itself: Wrap a cinder block with a furnace filter. Use bungie cords to hold the filter in place. Drop the cinder block in the river at a spawning site. The sticky filter will catch some of the descending eggs. Check the cinder block every day. Take any eggs on the filter to a streamside rearing facility. Raise the hatched nmé frye (larvae) until fall, when they'll be released back into the river.[16]

Greg tells a story about summer interns violating egg collection protocols while he and Kathy were filming. The interns were pulling their cinder blocks out of the water to pick off the eggs in the canoe. Greg and Kathy observed cinder blocks out of the water for upwards of 45 minutes, which would kill the eggs. This explained why supervising biologists were getting no viability. After the hubbub from this incident, the team started using coolers to pull out cinder blocks, hence keeping the eggs in water.

The streamside rearing facility was another important innovation from the LRBOI.[17] The idea was to keep nmé frye in their natal water, which had a unique geochemical signature. Ideally the growing fish would imprint on the Manistee River.[18] Nmé would thereby have a biophysical sense of place akin to the deep cultural sense of place among the Anishinaabe. This would be a revival of the group's "cultural and ecological connection to their nonhuman kin," as expressed by Marty Holtgren and Stephanie Ogren (and also Kyle Whyte), the tribe's first nmé biologists.[19] The trio wrote hopefully for a renewed "system of symbiotic living" between nmé and Anishinaabe peoples, a "lasting connection . . . between fish and human."[20] "By putting our hands in the water and having the young nmé slowly swim away," said Holtgren, "we are seeing ourselves in a different way."[21]

When Gregory A. D. filmed *Manistee Nmé*, the LRBOI fishery innovations were already a norm for Great Lakes sturgeon restoration. It was wonderful working with the Little River Band of Ottawa Indians, and later with the Potawatomi, says

Greg. "Our favorite so far," Kathy agrees. The Manistee nmé population is tiny compared to that of the St. Clair River. But she and Greg found a six-foot nmé in the Big Manistee, and filmed it in high definition. "Magical, breathtaking, an honor and a privilege," Kathy told the *Manistee News Advocate*.[22]

Even before they began filming, the two experienced a revelation during a planning meeting with tribal leaders. Here's Kathy's stream of consciousness:

"Never once did anybody use the word 'harvest.' That is always the ultimate goal of the dominant population, to get it so fishermen can catch 'em, so people can eat 'em, that's always what it's about.

"Not just because you need to have sturgeon in the Great Lakes, just because they're supposed to be there, and they're part of the habitat and the ecosystem, and I don't know why God put 'em in there or whoever did put 'em in there, the Great Spirit of the Turtle or whoever did it, but they're just supposed to be there, and I don't need to know why, I just want 'em out there, and not because I need to eat them.

"So that's how the tribe is," she closes. "They were strictly about, let's have the sturgeon back just because we want sturgeon in the river. And if we never see the results in our lifetime, we're okay with that. If our grandchildren know we tried, that's worth it. We're doing this so they know we tried."

An Interview about Seeing

Their near lifetime of seeing what most of us can't brought me to the Gregory A. D. home and studio, an old, converted country church in Lakeport, Michigan, just north of Port Huron. I had to make peace with the sensory overload of their place. A collection of naked mannequins was especially distracting. But eventually we got to the heart of things and considered the following:

With respect to the waters that surround and connect us, most humans have great *absences* of sensory experience. Sight, touch, smell, taste, hearing, and now a sixth sense, "proprioception"—the physicality and positionality of motion, a fluid sense of our embodiment in space: Without our sensory involvement, Lake Michigan is, actually, alien, at least to most of us, as in, we're alienated from its world below. The same for Greg's beloved St. Clair River, home and habitat to him, absent and alien to me.

What is it like to swim with a six-foot lake sturgeon? To say a neighborly hello to a bristling male round goby glaring at you from the entrance to his Budweiser can, daring you like a Chihuahua to threaten home and family? To watch a big not-hungry walleye hanging out near the seawall, taking a little breather from the current, inches away from some hopeful fisher's bait?

To discover the largest lake sturgeon spawning site in the Great Lakes, deep, deep in the St. Clair River, a place that even sturgeon scientists and conservationists hadn't known about?

What's it like to explore the same river for fifty years, as Greg has, so that he knows the river through time and space the way farmers know every changing contour of their fields and woods? What difference does it make that he and Kathy can see the rivers and freshwater seas of the Great Lakes in ways most of us never will?

But before more profundity, I still want to know about "bad divers."

River People

One day Greg went to the river to make a dive, and a group of anglers ran up to him. "Dude, can you come over and dive here for us? We think somebody did something right THERE," pointing to the water, "because six of us lost our tackle last night." Greg dove down, and sure enough, he found at least ten different traps set to snag fishing lures. Propping up a metal rod will snag them, he explains, or a tree branch, a pile of rocks, a window screen. "Bed springs was always something that people would throw in to catch all the lures." The window screen is easy pickings, says Kathy. "This piece of metal's loaded with fifty dollars' worth of tackle that he [bad diver] plans to bring up and sell right back to the guys he just took it from!"

When Greg and Kathy first started diving off the seawall at Pine Grove Park, the anglers would turn their backs and shun them. "For the last fifteen years," Greg says, "I've worked very hard to change their perspective."

Complaints about scuba divers run a gamut. They scare away the fish, mess with gear, steal tackle, cut lines. That makes them assholes, and on top of that, they're snooty. Not true about divers scaring fish, Kathy laughs. Or maybe hijinks by one person became an urban legend generalized to "scuba divers." And yeah, "it's kind of social, cultural," says Greg of the antagonism. Shore fishers are more likely to have old beater trucks, to smoke and drink in public. They're "that rough, tough guy," as Kathy puts it. Whereas diving "is a disposable income sport, and it's not cheap." So there's an assumption on both sides about class, income, attitude.

Anglers and divers share the water, but they don't mingle. The result, says Greg, is that "it only takes one diver to go down and cut all their lines." Then they'll hate divers for life. He recounted the time an angler who hadn't seen Greg before threatened to slit his throat if he dove there again. "Dude!" Greg flinched, "it wasn't me!" The other men at the wall rushed over. "Whoa, this is Greg, Greg's cool, he's the good guy."

They encountered hostility on the Big Manistee River, too. The Big Manistee is celebrated for its fly fishing. King and coho salmon and steelhead have major runs on the river. Before the *Manistee Nmé* documentary, Kathy doubted friends who told her that in one place anglers line the river bank shoulder to shoulder. "I didn't believe it," she says, still impressed, "and honest to god, they are shoulder to shoulder, and they cast in order, so the first guy casts, and as soon as his is clear, the next guy casts, and his is clear, the next guy, so that they don't all tangle up on each other."

On Kathy and Greg's first day to film in the river, men spotted the Gregory A. D. van, and accosted them in the parking lot. "You don't think you're diving here, do you?" Down at the boat ramp a larger group warned, "You can go back and get a park ranger, but he shouldn't bother coming unless he brings two sheriff deputies or a couple of state cops. Otherwise you're not getting into the water here with a scuba tank."

Okay, never mind. Retreat. "We came back with a boat, did the center of the river."

In Port Huron, Greg's status as a "good diver" also began in a parking lot. One day a large man approached him as he got out of the van. "He's real gruff, real big and gruff," Greg describes. "You'd think he'd be a mean person, but he turned out to be pretty nice." Kevin wanted Greg to go in the water right away. He had a salmon on the line, but it was tangled on a pipe.

"It's hooked on a pipe right there, and you could see it, and the fish is still on it, and could I just go untangle that? I got my tank on, went in, got the fish untangled from the pipe, let it go, he got the fish."

After that, Greg became shore fishers' eyes and hands underwater. "If they've had a lure that they've caught so many fish on, they've had it for twenty years, and it's on the bottom now, they basically want to kiss me when I come up." Absolutely, Kathy says, they'll even call: "Can you go in tomorrow? Is there any way? Is Greg free?"

Greg's fishing buddies were always curious about "down there." How many fish? Are you scaring the fish? Did you get my lure? I haven't had a bite in eight hours, so no walleye down there, right?

Yes, there were walleye. The walleye came over. The bass came over, too.

Oof.

The men watched out for Greg, kept track of his time below. He freaked them out one day when he stayed under a lot longer than usual. He'd been lying on a fish nest in a celestial zen state, communing with the life swirling around him, mind like water, as the Japanese metaphor goes. Some divers get thirty minutes from an air tank. Greg could go more than two hours by "sipping" the air in his tank. When he came up everyone got in his face. "Don't do that to us again!"

Greg once found something like a carpet that had suddenly materialized in the riverbed. "Things get buried and unburied," he notes, and this had probably been buried for years. But the emergent thing wreaked havoc on the fishing. All of that hooking and pulling exposed more and more of the mystery textile. People "were losing lures like crazy," Greg says, still amused. "There were just lures all over, and I was gettin 'em." Kathy laughs again. "They wanted it out bad," referring to the anglers.

"So finally," Greg resumes, "I felt real guilty [about collecting so many lures], so I went down, and got a rope around it, and it was heavy, a big piece of carpet basically, and got the line on it, had some fishermen up there, gave 'em a tug, and I filmed the whole thing. Might as well get something out of it. They couldn't even budge it off the ground, so I had to get under there, and pick it up, and get the suction out of it. Finally it came out. I would get under it and hold it and take it to the wall." Big Ed and the rest of the group pulled it up and out.

The water-logged carpet was actually a twelve-foot remnant of felt from the factory machinery of Dunn Paper. Dunn was one of two pulp and paper mills in Port Huron. Most likely a worker had thrown the worn-out felt into the St. Clair, where it drifted downstream, and eventually got buried.

There was an economic duality to Greg's diving. On one side was the more detached business transactions between Gregory A. D. and its clients. On the other side was Greg's place within a community of river people. The business involved things people lost in the water at marinas or docks or while boating. Gregory A. D. charged a flat $100 a dive to find keys, jewelry, whatever.

Typical clients were people who forgot about glasses in a shirt pocket. They would lean over the water, then PLUNK. A local mortician kept losing the same pair of prescription glasses at his dock. After the third time his daughter scolded, "You called Greg again? You're not paying him enough!" He sheepishly paid Greg $200 instead of the usual $100.

There was the little boy with his father in their fishing boat. Dad handed over the expensive rod so his son could reel in a big fish. Naturally the fish pulled the rod from the boy's hands. To their joy later, Greg located the rod on the river bottom. That was a great day: $100 and "another fisherman out there who will always think highly of divers."

They had a comical moment with a woman who lost her wedding ring off a dock. Seconds after he went down Greg broke the surface waving his hand and crying, "I can't do it, it's too scary down there."

What the?? "Get your ass in that water! Go back down there and find my goddamned ring!"

Man and metal twinkled. Greg had the ring on his little finger as he waved.

The hundred-dollar dives were different than taking out the Dunn felt. The felt was bound up in an actual community with its own economic and social fabric. The fishing tackle Greg collected from the river reveals this other dynamic. He kept a hanging board in his van with merchandise. Whereas a new lead sinker ran five to six dollars, Greg sold those he collected from a dive for $1.50. A new lure might run six to nine dollars. Greg sold them for one to three dollars. Once Kathy needed gas money, so he "hit a couple of parking lots and brought home fifty bucks." In the summer, recycled tackle brought them anywhere between $1,500–$2,500.

One dollar lures were not the whole of it, however. Kathy points to Greg's loyal following. The anglers trust him enough that they'll bring shore-fishing newbies over to the Gregory A. D. van. "If you need anything he's your guy," they advise. "I've tried to work at it," Greg says of the warm relationships, "because my favorite area is the parking lot where they do all their fishing."

In his own words, the parking lot scene unfolds like this:

I'll be in usually for about an hour and a half, and I'll come up the ladder, and I've got a down line. I can put a jug on there with all the fishing tackle. And now all's I do is, I come up the ladder, and I walk to the van, and everybody comes . . . and I just walk over to the van talkin', and I'm putting my gear away, and my crew goes over, pulls up all the sinkers [sometimes fifty or sixty pounds of lead].[1] This last time George, him and somebody else, were trying to pull this up over the wall for me, and they finally get it up, and George went, "How the hell do you drag that around down there?" I didn't say anything. But all's I do, I go over to the van, they haul it up, they get my flag [the diving flag from the wall], and they come over to the van. They sort it all, they take all the clips off, they cut the lines. People buy stuff from me. Then they go and hang [the rest] on the board—two dollars, one dollar.[2]

This micro-economy is ecological as well as social. Greg describes how he removes contaminants from the river when he dives. "In my bucket, or even just in my pockets, I'll bring out all the rusty batteries, the flashlight batteries...and all the fishermen think that's fantastic because it's not going to leak." ("And they're eating the fish," Kathy points out). "If I ever see any battery, I just make sure I get it out, and so they love me for that."

Worse yet is the tangle of tackle that no one but he and Kathy see. People don't think about the consequences of lures with hooks, which can still snag live fish. "Every year we find multiple fish snagged on the bottom," laments Kathy. That's a wasted kill, and it's a fish that has to suffer because it has to die on the line."

"It's the diving ducks that you really gotta worry about," Greg adds. "They'll go down and get tangled in the stuff. Most divers, they'll come along and see a lure, and they'll just take the lure or take the sinker, and they'll leave all the fishing line." ("And leave all the filament," says Kathy). "Well, sometimes there's thirty foot of line down there." Multiply this gear across the St. Clair and Detroit Rivers, or Lakes Huron, St. Clair, and Erie, across the Great Lakes, and eventually across the world's oceans, and Greg and Kathy's firsthand experience points to a global danger. *National Geographic* recently reported on a new study of the Great Pacific Garbage Patch, believed to be "the world's largest collection of floating trash."[3]

The study debunked plastic bottles as the patch's principal component and culprit (though plastic bottles were not innocent). The majority of the Patch consists of fishing gear.

In the ocean, abandoned "ghostnets" in the Patch and beyond snag or strangle dolphins, whales, seals, turtles. In the St. Clair River, lost and abandoned gear snags water birds, especially diving birds like mergansers. "They can't bite through it" (Greg and Kathy alternating), "they can't cut it, they can't get out of it. Crayfish and diving ducks, they're trapped and they're gonna die." That's why, Greg finishes, "when I find a sinker and it's got a line, I sit there and wind, wind, wind, wind—sometimes you think it's never going to end, cause there's thirty, forty feet of it."

———

There's still a question nagging at the skeptical me. Good will aside, was Greg really part of a "community" at the seawall, the parking lots, Pine Grove Park? To be sure he and Kathy knew lots of anglers. But what if it was random motion in those places? Just fishers, bicyclists, skateboarders, dog walkers, divers, individually enjoying their transient moment on the water? What is community anyway?

"What Is Community Anyway?" is the title of an article from the *Stanford Social Innovation Review*, published by Stanford University's Center on Philanthropy and Civil Society.[4] And here's an earlier article by cultural anthropologist Kathleen M. MacQueen and her associates ("et al.") asking the same question: "What Is Community? An Evidence-Based Definition for Participatory Public Health."[5] The authors agree that community is a slippery concept. Yet Greg and Kathy definitely saw a community of river folk—fluid, seasonal, informal, mostly male, but a real community, nonetheless. Greg saw himself as one of its members. I, on the other hand, saw a time lapse of random people moving in and out of view.

I was wrong, though. I wasn't seeing. The MacQueen article shows this.

MacQueen's research examined whether one can define community, and more specifically, "identify core dimensions of community." The researchers carried out interviews with four entirely different groups of people in four different places. Each group represented a public health initiative requiring community-level outreach—African Americans in North Carolina, gay men in San Francisco, injection drug users in Philadelphia, and HIV vaccine scientists nationwide.[6] Interviewers prefaced their questions with: "The word 'community' means different things to different people." Then they asked interviewees: "What does the word community mean to you? What is a community?"[7] Three core elements of community stood out across all the groups (out of five core elements identified in the study).[8] Abbreviating from MacQueen et al., below were the three core elements and a few follow-up descriptors:

Three Core Elements of Community

- *Locus, a sense of place, encompasses:*
  - ✔ a locale, boundaries, something that could be located and described
  - ✔ specific areas, specific settings
  - ✔ one can be physically in, even if not a member of, the community
- *Joint action, a source of cohesion and identity, involves:*
  - ✔ socializing, hanging out, conversing, intermingling, shooting the shit
  - ✔ keeping people informed about what's happening
  - ✔ watching over, checking up on, looking out for, keeping an eye on each other
  - ✔ doing things together, acting together, participating, planning, getting things done
- *Social ties, the foundation for community, encompass:*
  - ✔ interpersonal relationships between members and leaders
  - ✔ whom they trust, with whom they feel comfortable
  - ✔ who care about each other, with whom they interact, hang out, connect
  - ✔ who are known to them, who are seen in the background

Adapted from MacQueen et al., 2001

When it came to Greg and anglers at the seawall, ✓ for a sense of place, ✓ for joint action, ✓ for social ties, multiple ✓s for the descriptors.

Community is no small matter. This is as true for water conservation as it is for public health, and one dimension of water *is* public health. A community creates opportunities to confront problems, such as the Dunn felt. A community can effect a bigger change, too, if that's what people in the community want. Hence all the academic and professional interest in "community studies," "social innovation," "capacity building," and the like.[9] Gregory A. D. brought to St. Clair River communities the power of seeing underwater. Seeing became a catalytic change agent.

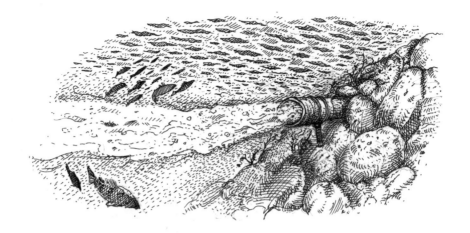

Power in the Visual

In the fall of 2010, shore anglers recruited Gregory A. D. to help with a longstanding complaint against Domtar, a Canadian company with a pulp and paper mill in Port Huron. Their complaint eventually included Dunn Paper. Fishers know these waters, Kathy prefaces the story, "they know everything that's going on down there but they never see it." With the Domtar situation, being unable to see was a problem.

Both Domtar and its Port Huron mill had a deep history in timber and forest products.[1] Domtar went back to 1848 England and its start as Burt, Boulton Holding Ltd. Founder Henry Potter Burt had developed an anti-decay treatment for raw lumber. This was a farsighted invention. An American railroad boom would soon demand enormous quantities of wood for railroad ties, and Burt-Boulton's treated lumber was more cost effective over time. The company grew, diversified, and in the early twentieth century moved to Montreal, Quebec, as the Dominion Tar and Chemical Company.[2] North American forests and other vast quantities of natural resources were close by, as was the surging pulp and paper industry in the Northeast and Great Lakes. In 1965, the company shortened and softened its name to Domtar, and by the 1990s, pulp and paper became Domtar's "core business."

Domtar's Port Huron mill came on line in 1888 as the Michigan Sulphite Fibre Company. In 1916 the company merged with Port Huron Paper to become the Port Huron Sulphite and Paper Company. Fast forward to the 1980s and an era of instability for the North American paper industry. Paper making technology and consumer markets were changing. Older mills were outdated. A globalizing industry was abandoning historic "paper cities" for new territory. Transnational mergers and acquisitions became the norm.

In 1983, international conglomerate Pentair Inc. bought Port Huron Sulphite and Paper. Pentair saw this acquisition as "another debt-laden but revenue-heavy paper mill."[3] Canadian company E. B. Eddy bought the mill from Pentair in 1987, then Domtar acquired E. B. Eddy in 1998. In that era, mill closures were dreaded but common events across the Great Lakes. By contrast, the Port Huron mill survived until 2021. Before its own closure, Domtar's Paper Made Here promotional materials proudly reported a Port Huron workforce of 250 that annually produced up to 114,000 tons of lightweight and specialty papers and packaging.

Domtar's mill was on the Black River, north of where that river meets the St. Clair River. But the company did not run its wastewater to the Black River. Instead, it ran discharge pipes a mile through Port Huron to an outflow into the St. Clair River. Greg had ideas about why the company used the St. Clair for discharge. Perhaps the slow current of the Black River meant that waste would stagnate in the still water? Perhaps mill owners once subscribed to a now archaic twentieth-century notion, that "the solution to pollution is dilution?"

History supports both propositions.[4] The Black River ran thick and putrid in the mill's first decades. In 1890 a Port Huron fisherman decried the pollution. "The Fibre Works . . . let their acids into the river and poison the water and kill the fish by the thousand, and the stench from the water is simply intolerable."[5] He was speaking to S. C. Palmer, statistical agent for the Michigan Board of Fish Commissioners. "I think the commission ought to do something about it if it is possible," he told Palmer. Palmer also talked with a Captain Moffat, who reported acids "so strong that they destroyed the paint on his boats, so he had to paint them below the water line several times last summer because of it."[6]

More than one hundred years later, any mill waste was mostly invisible. Domtar's outflow pipe was anchored to the St. Clair River bottom, out of sight. This same spot was a favorite for shore fishing, not just because of walleye, but also because it was accessible to people with physical disabilities who wanted to fish.[7] For years, Greg says, "fishermen complained that all this gunk was coming down, getting on their lines." They would pull up slabs and strings of a sticky substance that also glazed their rods, lines, and lures with its gross gray film. Grass on the river bank was slimed, too.

A slimy spot of shore near a paper mill outflow . . . not quite a sun-dappled image of nature communing. But nature communing such places can provide. In *Body of Water*, poet and fishing guide Chris Dombrowski contemplates his hometown river near Lansing, Michigan. Locals disparaged the Red Cedar River as the Red Sewer "for the abundance of shopping carts, diapers, and condoms found peppering its banks." But the same river was where a friend taught a teenaged Dombrowski "how to catch smallmouth bass, largemouth bass, rock bass, walleye, pike, a tiger muskellunge, Skamania steelhead, even salmon."[8] "And though I had no idea back then," he reflects, "I had witnessed the fusing of two seemingly unconnected worlds, the vile and the pure, and the sacredness of a place."[9]

Both Domtar and the state of Michigan denied anything more than an aesthetic problem of "visual fiber masses."[10] They insisted the mill had complied with its discharge permits for total suspended solids, including mandatory, albeit self-reported, testing for water quality.[11] The upshot: nothing to worry about.

Fishers asked Greg and Kathy, "Underwater, can you see it, the polluted zone?"

Yes, they answered. It was so bad they wouldn't dive near that pipe spewing its nasty shit.

Fishers again: Would Gregory A. D. go down and take a look? Maybe film it? Sure.

Greg and Kathy began filming upstream of the Domtar outflow. Upstream the water looked beautiful; fish were swimming. As they moved downstream toward the pipe they saw "a white plume coming out of it, with chunks." What did the plume look like? "Different colors of 'milk' coming out," Greg replies. "But then there's the particles, and some of 'em were bluish in color, I remember, and it's just kind of mucky, but it almost looks like material that's disintegrated into muck, so I'd put my hand behind it, and sometimes the pieces would be almost as big as my hand. And I'd be filming it as it's just floating downriver, and then I'd turn, and there's particles coming all over."

During filming they found a shopping cart on the river bottom, just downstream from the pipe. Greg filmed the cart multiple times and Kathy created a time lapse as it became covered with Domtar's visual fiber masses.

Sport fishers circulated Gregory A. D.'s footage among themselves, and Greg and Kathy released it to the St. Clair River BPAC.

What is a BPAC?

———

The 1987 Great Lakes Water Quality Agreement between the United States and Canada lists the St. Clair River as an "Area of Concern," or AOC.[12] From the original agreement of 1972 to its latest iteration, this has been the legal mechanism for cleaning up and restoring Great Lakes waters to "chemical, physical, and biological integrity."[13] A priority water, or area of concern, according to the agreement, "means a geographic area that fails to meet the general or specific objectives of the Agreement where such failure has caused or is likely to cause impairment of beneficial use or of the area's ability to support aquatic life."[14]

The Environmental Protection Agency coordinates binational areas of concern for the United States, while Environment and Climate Change Canada coordinates the same areas for Canada.[15] The St. Clair River Area of Concern has a four-way administrative agreement among the US Environmental Protection Agency (EPA), Environment and Climate Change Canada (ECCC), the Michigan

Department of Environment, Great Lakes, and Energy (EGLE), and Ontario's Ministry of the Environment and Climate Change. The quartet oversees a Remedial Action Plan for the St. Clair River. (See the glossary below, since agency names change, and their acronyms can look like an alphabet soup.)

Areas of concern leverage community and stakeholder involvement through their public advisory councils (PAC) or binational public advisory councils (BPAC). The St. Clair River BPAC includes members of the Bkejwanong (formerly Walpole Island) and Aamjiwnaang First Nations. For more than a century, the Bkejwanong fishery has been the proverbial canary in the coal mine, or the proverbial first responder, for environmental emergencies in the St. Clair River and Lake St. Clair. Representatives for sport fishing are also energetic members, no surprise, since their $7 billion Great Lakes industry depends on beneficial uses. Council members have a lot of advising to do. For the St. Clair River, their mission is to:

> advise the RAP Team on key aspects of the Remedial Action Plan Preparation and Adoption. This includes: the goals of the plan, problems to be addressed, planning methodology, public involvement program, technical data, remedial action alternatives, planning recommendations and adoption, plan implementation, plan funding and methods of enforcement.

The council's advice is not mere window dressing either. "The goal of all concerned," its mission statement concludes, "should be to arrive at planned recommendations upon which the RAP Team and the Advisory Council agree, and for which there is broad public support." Everyone's ultimate destination is delisting the area of concern by removing each impairment. The process can take decades, with slowdowns, holdups, and barriers along the way.

The St. Clair River BPAC responded quickly to Gregory A. D.'s film. US co-chair Janice Littlefield called a meeting with representatives of Domtar and the EPA. Domtar reps stressed their concern for the environment, while Rose Ellison of the EPA reiterated Domtar's compliance with its discharge permits. Littlefield then told attendees, "Now let's watch a video. Can you please push play? This was shot by a local diver last week."

There was silence afterward. The visuals were undeniable. Everyone saw that Domtar was not complying with its permits. Worse yet, what if the mill was contributing to the St. Clair River's beneficial use impairments (BUIs)?

To see why Domtar's transgression was a big deal, some context is in order.

Glossary of Acronyms

AOC	Area of Concern
ASEP	Atmosphere–surface exchangeable pollutant
BUI	Beneficial Use Impairment
BPAC	Binational Public Advisory Council (when the AOC is binational)
COC	Chemical of Concern
CRIC	Canadian RAP Implementation Committee
CSO	Combined Sewer Overflow
ECA	Environmental Compliance Approval (Canadian)
ECR	Environmental Compliance Reports (Canadian)
EPA	United States Environmental Protection Agency
EC	Environment Canada (today Environment and Climate Change Canada)
ECCC	Environment and Climate Change Canada
FCSV	Fish Consumption Screening Values
FOSCR	Friends of the St. Clair River (Canadian, with an American counterpart of the same name)
GLWQA	Great Lakes Water Quality Agreement
GLSCF	Great Lakes Sport-Caught Fish
MDEQ	Michigan Department of Environmental Quality (today MEGLE)
MEGLE	Michigan Environment, Great Lakes, and Energy (most recent as of 2020)
MDNR	Michigan Department of Natural Resources (later merged with MDEQ, then separated again)
MDNRE	Michigan Department of Natural Resources and Environment (when MDEQ & MDNR merged)
MFCAP	Michigan Fish Consumption Advisory Program
MISA	Municipal/Industrial Strategy for Abatement (Canadian)
MWRA	Michigan Waters Resources Act
NPDES	National Pollutant Discharge Elimination System (US)
OMOE	Ontario Ministry of the Environment
PAC	Public Advisory Council (when the AOC lies within one country)
PCB	Polychlorinated Biphenyl
PDF	Portable Document Format
POP	Persistent Organic Pollutant
RAP	Remedial Action Plan
RRAI	Rouge River Aesthetic Index
SCR-AOC	St. Clair River Area of Concern
SCRCA	St. Clair Region Conservation Authority (Canadian)
SLEA	Sarnia–Lambton Environmental Association of member industries (Canadian)
TSS	Total Suspended Solids
WIFN	Walpole Island First Nation (today Bkejwanong First Nation)
WWT	Wastewater Treatment
WWTP	Wastewater Treatment Plant
WPCP	Water Pollution Control Plants (Canadian)

Industry-based environmental sustainability abbreviations used by Domtar:

EMS	Environmental Management System
FSC	Forest Stewardship Council
ISO	International Organization for Standardization
ISO 14001	Voluntary technical standards for an effective EMS
ISO 9001	Voluntary technical standards for a QMS
PEFC	Programme for the Endorsement of Forest Certification
QMS	Quality Management System
SFI	Sustainable Forestry Initiative
UL ECOLOGO	Underwriters Laboratories' standards for reduced environmental impact

Note: Tracking the life cycle of an area of concern means reading and writing in capitalized shorthand, like this specimen from an agency document: "DNRE AOC coordinators work closely with the local PACs to populate the AOC Action Tracking Table with actions that are required to advance the restoration of BUIs for each AOC."

The non-fluent are at a disadvantage with such acronymic pidgin languages. Hence this partial glossary for citizens who wish to explore historical SCR-AOC agency and industry documents.

Compiled by Lynne Heasley

A Not-So-Objective Introduction to the Fish Consumption Advisory

As of 2010, only one beneficial use impairment (BUI) had been removed for the St. Clair River, "restriction on dredging activities," and this removal was for the US side only. The restriction remained for the Canadian side. This had been a BUI because dredging would have stirred up toxic contaminants in river sediment—mercury, dioxins, chlorinated benzenes, polychlorinated biphenyls, and others. Restriction on dredging activities was a foundational BUI in areas of concern that had a history of paper manufacturing. Paper mills used huge quantities of water. That's why the industry arose along major waterways. For much of the twentieth century, mills also relied on public waters for free waste disposal, as did all heavy industry concentrated on water frontage.

In the St. Clair River, paper mills were not the only contributor to toxic sediment. Petrochemical facilities like Dow Chemical of Canada loomed larger, especially on the Canadian side, in Sarnia's Chemical Valley. From 1949 until 1970, Dow Canada—by itself—spilled an average of thirty pounds of mercury per day from two mercury cell chlor-alkali plants.[1] Chlor-alkali plants produced chlorine for bleaching paper.[2] Dow admitted to spilling seventy-five pounds per day from July 1969 until February 1970, when the Ontario Water Resources Commission took enforcement action.[3] Dow's total mercury releases into the river may well have reached three hundred tons. When the commission tested river and lake walleye, the fish had mercury in their tissues at such high levels that Ontario closed the commercial fishery from the St. Clair River to Lake Erie.[4] On the US side, Michigan closed the same corridor to all fishing. Walleye at that time, according to Ontario and Michigan, were unsafe to eat in any quantity. Lasting ten years, the fishing ban ravaged the Bkejwanong First Nation, a people at the mouth of the St. Clair River who depended on their fishery culturally, economically, and for family-level food security.[5]

There was also Dow's part in the monster "BLOB." "That one probably got more press than just about any release that ever occurred on the river," BPAC member Fred Kemp told *Lansing State Journal* reporter Beth LeBlanc for a retrospective of the event.[6] In August 1985, Dow Canada spilled 528 gallons of the dry-cleaning solvent perchlorethylene. The following November, scuba divers described big blobs with texture like tar in a 30,000 square-feet area of riverbed near the spill. Most likely, the perchlorethylene had aggregated other chemicals already in the sediment into a toxic playdough. "It's a solvent, so that's what it does for a living," Kemp explained. "It doesn't readily mix with water. If you had a film of hydrocarbons on the river bottom, it would have loosened them and released them, and then it would have collected in the low spots in the river where the velocities are lower."[7] Dow began vacuuming up the BLOB soon after the St. Clair River was named an area of concern.

Another scandal surfaced in August 2003, when the Northeast and Great Lakes experienced one of the largest power blackouts in American history. During the blackout, Sarnia's Royal Polymers Ltd. lost power to its cooling system, and in a series of unfortunate events, the plant spilled at least three hundred pounds of vinyl chloride into the river. In a criminal non-act, Royal Polymers did not report the spill for five days.[8] Close by were the communities of Wallaceburg, Ontario, and the Bkejwanong First Nation, both of which used the St. Clair River for their water supply. Thirteen Michigan drinking water treatment plants also lay downstream. Medical anthropologist Christianne Stephens was working with Bkejwanong communities when they learned that vinyl chloride had likely contaminated their water. "For residents, the phrase 'left in the dark' took on two meanings," she recalled, "the power outage itself and connotation of the 'masking' and purposeful obfuscation of knowledge by industry regarding the threat to human and ecosystem health."[9]

The most notorious spills were punctuation points in a chronicle of indifference, or of intentional brutality, as Bkejwanong communities saw it. Mercury, the Blob, vinyl chloride were traumatic wave crests, and between crests were rolling waves of anonymous spills and allowable releases. The paper industry contributed to the river's toxic load with its own unseen but dangerous chemical riptides, among them polychlorinated biphenyls, or PCBs.

PCBs were manmade industrial chemicals widely used for fifty years, until their ban in 1979.[10] Their chemical stability and insulating properties made PCBs ubiquitous in American manufacturing, from consumer products to factory equipment—appliances, paint, plastics, transformers, capacitors, electric systems, hydraulic systems. Paper mills also used PCBs to manufacture carbonless copy paper. Once unloosed in the world, the same chemical properties that made PCBs convenient for producing a piece of copy paper made PCBs a public menace.

PCBs were (and are) stable, so they would not break down for decades; they cycled through the local environment—soil, water, air—and they were light enough for great quantities to travel worldwide on wind currents, even to the Arctic.[11] A dangerous endocrine (hormonal) disruptor, PCBs also biomagnified up the food chain from sediment to plankton to small fish to large predator fish to humans, threatening the reproductive health of human and nonhuman communities alike. PCB risks reported by federal agencies included cancer, miscarriage, cognitive impairments, behavioral problems, birth defects, and damage to a fetus's or child's developing brain.[12] By dumping their PCB-laden waste, paper mills created toxic "hotspots" in their home rivers. The Kalamazoo River in Michigan and the Fox River in Wisconsin also became infamous AOCs because of papermaking.

As of 2010, the St. Clair River BUI "restrictions on fish and wildlife consumption" applied to species like walleye, smallmouth bass, and yellow perch, which had "elevated levels of mercury, PCBs, pesticides and mirex and photomirex."[13] Based on PCB and dioxin levels, the "Michigan Family Fish Consumption Guide" and the "Ontario Guide to Eating Sport Fish" advised people to eat no more than six meals of walleye per year.[14] Only the "Limited" and "Do Not Eat" meal categories were more restrictive. In 1970, Michigan had announced the first "Do Not Eat" advisory in the United States—for fish in the St. Clair River.[15] Soon fish consumption advisories opened a window on water pollution across the Great Lakes. They made an invisible and silent blitzkrieg of toxins visible to the fish-eating public.[16] Forty years later, advisories "have become so central to modern pollution regulation that they are essentially naturalized," says historian Nancy Langston. They are "here to stay" and "just a part of fishing the St. Clair," states the *Great Lakes Echo*.[17]

Because fish consumption advisories were grounded in laboratory testing, they had the benign appearance of objectivity. Yet critiques of fish consumption advisories bubbled up. Great Lakes fishing communities and researchers in those communities came to see objectivity and neutrality as illusions that hid grave injustices.

The advisories were based on the toxic load in tissues of an indicator species.[18] The "6 meals per year" category meant that walleye had between 0.21 and 0.43 parts per million (ppm) of PCBs in their tissue.[19] By contrast, walleye in the "16 meals per month category" had less than 0.01 ppm of PCBs in their tissues. At this low level, walleye could safely be part of someone's diet. Advisories also distinguished between healthy adults and vulnerable populations, including children, pregnant women, and people with chronic diseases.

What could be more useful and neutral to someone about to eat fish than a scientifically-based fish advisory? Deep down, however, advisories were not objective; they were very, very personal.[20] There were disconnects between the implied morality of advisory metrics (be fish-aware when you cook a fat walleye for your loved ones . . .), and moral reasoning within groups on Bkejwanong or at the seawall. Take one warning from the Michigan Fish Consumption Advisory Program:

Human fetuses are exposed during development to PCBs in contaminated fish that the mother eats. Exposure to newborn and older babies could occur through the mother's breast milk. In addition, infants may have a reduced capacity to metabolize and eliminate PCBs due to still developing organ systems. If toxic exposure levels are high enough during critical growth stages, the developing body systems of children can sustain permanent damage.[21]

With that dark omen, think about this one-two gut punch of a calculation: Eat a meal of large walleye from the St. Clair, and you knowingly add a dose of PCBs and dioxins to your body's fat stores. Breastfeed your baby, and you knowingly unload your own accumulated toxins into your baby's tiny body. In 2005, Florence Williams scared nursing mothers with her *New York Times* article, "Toxic Breastmilk?" in which she wrote, "When we nurse our babies . . . we also feed them, albeit in minuscule amounts, paint thinners, dry-cleaning fluids, wood preservatives, toilet deodorizers, cosmetic additives, gasoline byproducts, rocket fuel, termite poisons, fungicides and flame retardants."[22] Could Williams's sincere effort at raising awareness lead to problematic messages? How about something like: "Breastfeeding: It's on You, Not Them." Or: "Breastfeeding: Just Another Way to Be a Bad Mother."

Now banish the omen for a moment, and weigh the reasons to go ahead and eat that wild-caught walleye (or steelhead or salmon . . .). In North America, most exposure to PCBs and dioxins comes from meat and dairy, so avoiding fish will not spare you. Eating fish two or more times per week (the American Heart Association's recommendation) protects heart and brain, lowering your risk of heart attacks and strokes. Fish are one of the best sources of Omega-3 fatty acids, essential for a child's developing brain and nervous system. Conversely, a low level of Omega-3 fatty acids retards a child's development. Nutritional neuroscientist Dr. Joseph Hibbeln warns that Omega-3 deficiency also puts kids at risk for major depression.[23] Renowned food writer Michael Pollan worries that Omega-3s are "the vanishing nutrient" in the American diet.[24] When Harvard's T. H. Chan School of Public Health posed this question, "Fish: Friend or Foe?" it answered, "fears of contaminants make many unnecessarily shy away from fish."[25] As for breastfeeding, breastmilk is still a baby's finest food. Commercial baby formula cannot replicate its rich nutritional matrix and immune-system-building powers.[26]

So "what is a woman to do?" queries Nancy Langston in her case history of toxaphene in Lake Superior fish. "Does she eat those fatty fish . . . secure in the knowledge that their Omega-3 fatty acids are helping neurological connections form? Or does she shun them?"[27]

Bad omen or no, fish consumption advisories have an optimistic premise. Scientific information, together with outreach to target groups, should empower people to make their own choices. Even complex pros and cons fit within this self-deterministic model of individual decision-making. To eat or not to eat the fish? You decide.

But scale up from individuals to entire fishing communities, and fish consumption advisories can be another source of alienation—the opposite of their good intent.[28] If you're a member of a subsistence fishing group (i.e., poor people near water), or a fishing culture, like the Bkejwanong First Nation, fish are a mainstay in your weekly diet and native cuisine.[29] This is not an individual choice. Farming communities would no longer be farming communities if people didn't farm. Ranching communities would no longer be ranching communities without livestock. Likewise, fishing communities are not fishing communities if people don't fish. What if fish consumption advisories undermine a collective sense of place? Undermine water as a source of cohesion and identity? Undermine river and lake as a foundation for community? Without these elements of community, can the community itself survive?

For First Nations on the St. Clair River, this was not a safe, objective thought-exercise. They were living a nightmare. A look at the sustainability movement for regionally sourced foods shines light on this nightmare.

———

In the twenty-first century, regional food is in. Celebrity chefs go on vision quests for authentic regional cuisine. Visionary academics look to regional food systems, cultural foodways, and local food security for revitalizing hard-hit urban and rural places. One idea is that a vibrant regional food economy is a feature of resilient, flourishing communities. Another idea is that close connections to our food can bind us in a web of care, not just care for ourselves, but for our landscapes and waters. The late Michigan writer and food sage Jim Harrison reduced the philosophy and methods of regional food to their essence. "Eat where you live," he instructed.[30]

What people mean by regional food can vary. Usually they mean food from nearby agriculture, horticulture, and livestock—to wit, community supported agriculture, community gardens, urban agriculture, urban apiculture (that's honey bees), food forests, artisanal foods, farmers markets, farm-to-table networks. Jim Harrison himself once sang the praises of Clancy's Fancy Hot Sauce from Ann Arbor.[31] But in the human history of the Great Lakes, native fish are one of the original regional foods. They are a founding food. Regionally speaking, walleye, whitefish, yellow perch, and lake trout tower over Emperor Francis sweet cherries and Bell's Oberon.

Harrison would surely have agreed that fish are where we live. "Buy yourself a fishing pole," he once berated, "even now as I am dictating this in my auto I am heading back to my hidden cabin in the forest in the Upper Peninsula of Michigan, and my car is loaded with fishing tackle. I even like the less desirable species pike; I like perch, Lake Superior whitefish, lake trout, the small pinkish ones."[32] Harrison's food essays had the same lurid, earthy energy as his best-loved novels.

Harrison's life as well as his writing revolved around being in place and committing to nature. Place and nature made up his core. Sometimes Harrison looked around and shivered and raged at omens and nightmares. In "Cooking Your Life," he introduced his Michigan home as "usually a location of modest serenity."

> But there had been a recent photo of two cormorants with horribly twisted beaks, a mutation caused by the amount of toxins that industry continues to pour into the Great Lakes. I had to turn on the bed lamp to get rid of the cormorant image. I checked my notebook for a Lewis Shiner quote: "Pollution is simply alcoholism on a global scale. It grows out of the same kind of self-hatred, greed, and impatience, the same kind of delusions about control." There, that pinpoints it, and there was the image of the industries that border the Great Lakes, as reeling and drunken, raising their big, blotched asses and spewing chemicals into the once pristine lakes. It was difficult to be upset with Saddam Hussein for doing the equivalent on farther shores.[33]

Harrison brought words to a hard boil better than anyone. But the rest of us can still learn and practice. Fish consumption advisories for fishing communities? Sure, trim the fat off a fish to reduce toxins, but damnit, there's a war going on, and the Bkejwanong First Nation, and the Aamjiwnaang First Nation, whose reserve in Sarnia is surrounded by Chemical Valley, are in the middle of it!

The assault this time was not smallpox or rifles, or violent removal, or the Canadian residential boarding schools that stripped generations of Indigenous children from their families and culture and where brutality and abuse were epidemic. This time it was modern industry bombarding men, women, and children with toxins, waging "another form of genocide," as Aamjiwnaang activist Lindsay Gray charged.[34] It turns out that Canada's 1977 PCB ban grandfathered in PCB-dependent factory equipment. Companies in Chemical Valley could use such equipment until the end of service life, even decades more of service life. "Before it was TB and influenza killing us," said an elder from Bkejwanong. "Now it's pollution. Chemicals are the new smallpox blankets given to our people in the 21st century."[35] In this context, fish consumption advisories were Band-Aids stuck on centuries of traumatic injuries.

On the St. Clair River, the two First Nations faced a public health calamity. An unholy alliance of diseases plagued them. People died younger than their great-grandparents. Cancers were omnipresent.[36] Miscarriages and birth defects were a norm. Most shocking of all was the "disappearing boys." A 2003 study of Great Lakes fish consumers had already shown how maternal PCB exposure could skew a population's sex ratio of live births.[37] Higher loads of PCBs meant lower odds of having a male child.[38] The authors offered two plausible mechanisms through which PCBs might interfere with sex ratios:

- Reduced fecundity [fertility] "because of an increased rate of spontaneous abortion of male fetuses";

- Estrogen disruptions, or "estrogenic and anti-estrogenic actions of different PCB congeners."[39]

A separate scientific study of the Aamjiwnaang First Nation showed an abnormal shift in the community's sex ratio of live births from 1994 to 2003.[40] In her book *Everyday Exposure: Indigenous Mobilization and Environmental Justice in Canada's Chemical Valley*, Sarah Marie Wiebe compiled media shock-quotes following these studies: "lost boys," "birth dearth," "girl baby boom," "where the boys aren't," males "an endangered species."[41]

Wiebe argued that the "slow violence" of extreme toxic exposure demanded reproductive justice for Indigenous and other marginalized women of color, and that reproductive justice is a central component of environmental justice.[42] But rather than justice, she observed, Aamjiwnaang communities were subjected to "lifestyle blaming" for their diseases and reproductive problems.[43] Such shaming, she said, wrongly presumed "a rational, atomistic individual charged with mastering an unruly body" (a presumption shared by fish advisories), as opposed to a people living within the particular histories and geographies of an actual place, in this case the Aamjiwnaang Reserve.[44] In this place, a resident told Wiebe, "you can taste the toxins" in the fish. "You can smell 'em when you're cooking them."[45]

Wiebe also pointed out another perverse contradiction: that Aamjiwnaang "bear a disproportionate responsibility for proving toxic exposure and adverse health effects" before officialdom will admit their plight.[46] Indeed, the validity of the sex-ratio study became a dispute pitting First Nation against Province of Ontario scientists, who argued that the study's sample size of births on the reserve was too small for a valid statistical analysis, let alone conclusions about lost boys. And moreover, that sex ratios of births at the county level (rather than the reserve's small area within the county) were normal and mirrored provincial sex ratios.[47]

Wiebe zeroed in on two problems with scientific skepticism about Aamjiwnaang sex ratios: First, Lambton Community Health Services, which carried out one of the rebuttal studies, was not a neutral arbiter. The agency was defensive about Sarnia's image as an unhealthy place to live, as "Canada's cancer capital."[48] Second, a scientific paradigm in which, as Wiebe said, "such a small population cannot generate credible statistical results," put the Amjiwnaang in a no-win situation. "The biomedical power and authority embedded within the language of science, epidemiology, and statistical significance overlooks and discredits Aamjiwnaang's situated, experiential claims," she explained.[49] Nevertheless, when Bkejwanong played Aamjiwnaang in co-ed youth softball, the Aamjiwnaang team was all girls.[50]

Grieving was another dimension that big science couldn't capture. Aamjiwnaang and Bkejwanong communities had to process their personal and collective grief about toxic chemicals infiltrating their bodies, distorting their world. In her dissertation research, which was a partnership with Bkejwanong residents,

Christianne Stephens called this processing "Toxic Talk." Toxic Talk consisted of narrative discourses—webs, layers, fragments—a kind of existential accounting of lived experiences.

But don't mistake the grief of Toxic Talk for defeatism. Novelist and scholar David Treuer rejects the subliminal, reflexive framework many people apply to North American Indians. "We" are not perpetually tragic, points out Treuer, who is a member of the Leech Lake Band of Ojibwe in Minnesota. To the contrary, he says, we're still here, active shapers of the present and future. Treuer's sensitive but startling and searing *The Heartbeat of Wounded Knee* shatters intransigent over-romanticized stories that relegate Indigenous peoples to the past, or reduce them to tragedy.[51] He brings into sharp relief how one cannot understand America today without understanding Native America. To adapt Treuer's argument, one cannot understand North American conservation and environmental activism without understanding Indigenous conservation and environmental activism.

A proactive rather than reactive reading of Bkejwanong discourse is what geographer Emilie Cameron describes as "storying for change."[52] "It may be," says Cameron, "that it is precisely in small, local storytelling that political transformation becomes possible."[53] Bkejwanong and Aamjiwnaang First Nations exemplified the transformative energy of storytelling, or, as Cameron explains, "the capacity for stories to be practiced in place and to generate . . . change."[54]

Storying for change, Aamjiwnaang activists and water protectors joined climate activists on Toxic Tours of Chemical Valley. One person on the tour told *The National Observer*'s Fram Dinshaw that she was shocked to see the blighted Chemical Valley landscape. The tour, she said, "ends at the cemetery, which I thought was extremely poignant. . . ."[55]

Storying for change, the First Nations charged treaty violations: that Chemical Valley industries were trespassing on their sovereign treaty rights to fish and hunt by poisoning the fish and game; and that Canadian and Ontario agencies were complicit by refusing to punish illegal toxic releases.[56]

Storying for change, Bkejwanong and Aamjiwnaang communities bypassed big-data biomedical paradigms in favor of a powerful new paradigm, Two-Eyed Seeing.[57] In principle, two-eyed seeing was a decolonizing project in which Indigenous knowledge (one lens) and Western science (the other lens) became partners. In practice, these First Nations became trailblazers in participatory public health research.[58] Using citizen science approaches, they began testing and monitoring water quality. Using community-scale methodologies, they undertook community-mapping, bio-mapping, and body-mapping, all of which can reveal patterns in the collective experience of a community across different health problems.[59]

The ultimate storying for change is still aspirational: full "ecological citizenship."[60] While reaching for that most profound state of justice, though, Bkejwanong and Aamjiwnaang women remained active stewards of their own bodies, in tandem with their historical stewardship of water. They embodied a reproductive justice framework in which, according to Wiebe, "individuals must have the right and ability to reproduce in culturally appropriate ways."[61] Thus both communities helped shape new environmental policymaking and enforcement, including, and circling back to, the St. Clair River BPAC.

———

What does it mean to be so lovingly conjoined to a place that lies in the twisted toxic shadow of Chemical Valley?[62] To share all the lesions and illnesses between you, so that the survival of one depends on the survival of the other? And to live in symbiosis, so that the flourishing of one would allow the other to flourish? Also, what does it mean to know that your community could be grievously injured during any surgical separation from its homeland and home waters?

Of everything the Bkejwanong and Aamjiwnaang communities did, perhaps their most powerful storying, their most radical act of resistance and renewal, was this:

They stayed.

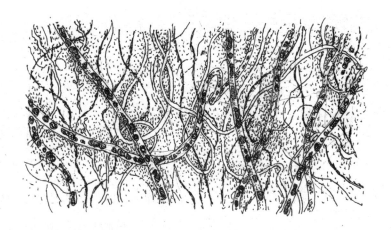

Rooted in Sustainability

Under such a shadow, removal of "restriction on dredging activities" from impairments on the US side of the St. Clair was a ray of hope. To clean up the river was to begin moving out of the shadow. But the same impairment remained on the Canadian side, and other impairments remained on the US side, including tainting of fish and wildlife flavor, degradation of aesthetics, degradation of benthos, loss of fish and wildlife habitat, and, of course, restrictions on fish and wildlife consumption. The St. Clair River BPAC wanted no holdup on those.[1]

Back in Port Huron, Domtar representatives had tried to swat away local complaints about their wastewater discharge. Now they had a credibility gap. Michigan Department of Natural Resources and Environment (MDNRE) spokesperson Mary Dettloff told the *Detroit News* that, "We're concerned because we've seen the same photos and videos and those images don't correspond with what the company is telling us."[2] Domtar had more at stake than this one pollution controversy. With the Gregory A. D. video, Domtar faced a reputational as well as an environmental crisis.

Domtar Corporation was burnishing its reputation for sustainability. The company had developed a line of Domtar EarthChoice office paper with the slogan "high quality paper with a conscience," proclaiming the line "quite possibly the most sustainable family of fine papers in North America."[3] Domtar designed EarthChoice, it said in its 2007 annual review, "for companies that are outspoken about their commitment to protecting the environment as well as on other environmentally sensitive issues."[4]

Domtar's 2009 *Sustainability Report* was titled "Crafting Paper Responsibly: Rooted in Sustainability from Forest to Print."[5] In the report, president and chief executive officer John D. Williams touted the company's sustainability commitments. "At Domtar," he said, "sustainability is an integral part of our activities. It guides our business and manufacturing practices throughout the life cycle of our paper, pulp and wood products, from the forest to our mills and all the way to our customers across North America and around the globe."[6] Williams also defined sustainability to include community well-being:

> Our commitment to our employees and our involvement in our communities are intrinsically linked to our success as a responsible company, in the same way as generating a solid financial performance for our investors and achieving governance and ethical business activities. As such, we must continue to build on the foundations laid over the years to contribute to a better world, one where future generations will reap the benefits of our decisions and actions today.[7]

One way that Domtar could implement best practices at its mills was through ISO 14001, the international standard that certified a company's effective environmental management system.[8] The Port Huron mill was certified.[9] What's more, from 2007 to 2009, the mill reported the lowest total suspended solids in the effluent of Domtar's "specialty paper mills" (which were among of the lowest of Domtar mills regardless of category).[10] One might treat this metric skeptically after seeing the plume entering the St. Clair River.

Now the company was bound by an interlocking chain of twenty-first-century expectations. First was a Great Lakes regulatory framework at state, federal, First Nation/tribal, and binational levels. This was buttressed by bipartisan political support to protect the Great Lakes. Second were emerging practices and rhetoric within the pulp and paper industry to meet a "triple bottom line" of environmental, economic, and social sustainability. In the United States and Europe, the industry proclaimed itself a manufacturing leader showing that conservation and sustainability could be profitable. Mills had made enormous strides in water conservation and recycling and in the use of less toxic, non-chlorine-based bleaching agents for paper. And third were local expectations for corporate citizenship. Domtar and Dunn were major employers whose jobs were critical to the local economy and whose products were a source of pride. But that didn't preclude community demands for good behavior, especially with water.

Domtar was blindsided by divers filming underwater. What Greg and Kathy could see and know firsthand and close up, they could also share through visual media, so that others could see and know, not firsthand, but definitely close up. On Michigan-sportsman.com, "wartfroggy" posted a smart, anticipatory comment in response to the Gregory A. D. footage. "The grey slime is pretty weird. It isn't paper, but looks like it could be bacterial growth that had peeled off from the inside of the discharge pipe. It would be interesting if they [Domtar] had it analyzed."[11]

No-nonsense about their impact, Greg says that, "we basically through our video changed what they do." Within weeks Domtar distributed a public brief.[12] Its sixteen bullet points flowed from acknowledging a problem—"Recently

Sample Domtar Talking Points on the Port Huron Mill Effluent

- We have a long tradition of cooperating and working closely with the DNR, DEQ and now DNRE to achieve our environmental obligations. All our permits are in place and are in compliance.
- Domtar understands the importance of water quality in the St. Clair River. Water is critical to the process of making paper hence the reason paper mills are typically built along rivers like the St. Clair River.
- The visual issue in our effluent is a substance that biologists call biofilm bacteria which are natural occurring in the Great lakes [sic]. Biofilm bacteria attach to tanks and pipes that store and transport water. Being very adaptive organisms, and given the warm atmosphere which our effluent provides, they grow fast and excrete a slimy material, making them resilient against efforts to control their growth. It is this biofilm that breaks loose or fluffs off from the effluent pipes when there are flow increases.
- We are in business to make paper and we are also committed to protecting the environment and listening to the concerns of our neighbors in the community.
- You can visit our corporate website at http://www.domtar.com/en /sustainability to review our position on environmental issues. And you may also use the email address (sustainability@domtar.com) to send your requests for further information.

From Domtar, 2010

Domtar Port Huron Mill was made aware of an increase in visual solids with our effluent"—through asserting good corporate citizenship, explaining the issue, laying out first steps toward a cleanup, and reaffirming its values of environmental health and sustainability.

In one bullet Domtar said, "Every day, mill water quality specialists gather and test samples of our reclaimed process water to make sure it complies with our MDNRE water permit."[13] This was the big discrepancy, the mill's clean bill of health when it tested its own water quality vs. the effluent polluting the St. Clair River and obvious to everyone at the site.

The early stages of Domtar's cleanup confirmed wartfroggy's hunch. Or maybe it wasn't just a hunch, maybe some kind of expertise peeked through the post. Domtar had tested its treated wastewater at the mill, not at the discharge point into the river. Unseen for years, that mile of pipeline had accumulated biofilm bacteria or "filamentous bacteria growth," as Domtar explained for a December 2010 BPAC meeting:

> Filamentous bacteria growth in effluent paths is a common and vexing concern among paper mills. These bacteria are present in the Mill's discharge from the clarifier to the lower outfall discharge point at the river. Biologists believe that paper fiber strands entangle in the bacteria growth giving the resulting mass the appearance of pieces of paper when changing water discharge velocities tear them away from either piping, the effluent wells, or outfalls. These light fiber masses follow the river currents, wrapping onto fishing lines and swirling in eddies at retaining walls downriver.[14]

Wastewater treatment involves microbiology, biochemistry, large infrastructure, and adaptive treatment protocols. Applied scientists at wastewater treatment plants even call themselves "bug farmers," since they cultivate the microbes that digest organic matter. But in paper mills, the system is often on the brink of a wicked problem, because of the enormous quantities of cellulose-dense organic forest matter to dispose after paper is fabricated. In pulp and paper manufacturing, the result is a complex, evolving, precarious system, an "organic machine," to lift historian Richard White's term.[15]

———

A quick author's interjection and detour here: I miscalculated by asking an organic chemist friend, who knows a bit about wastewater treatment, what level of detail that he, as a reader, might want when it came to Domtar's slime. In 2010, Domtar's immediate challenge was to reduce biofilm and fiber released into the St. Clair River. Would an organic chemist familiar with wastewater want details from an early point in Domtar's system? Perhaps starting with the mill's clarifier, which was a settling tank for removing solids? Or maybe, since plant design was part of Domtar's challenge, he'd be satisfied with an infrastructure sketch highlighting the manifold pipe from which smaller discharge pipes ran to their outflows at the St. Clair River? Or would he assume the ins and outs of infrastructure and prefer to bypass all that for real drama, the pollution controversy being negotiated among Domtar, BPAC, and the MDNRE?

Delicately, like one would hope of a professor with hundreds of pre-med students desperate to pass organic chemistry, Steve Bertman posed a few softball questions about the Domtar case (I thought I passed those questions with at least a "B+"):

Then abruptly, what permit was Domtar violating? (Its National Pollution Discharge Elimination System [NPDES] permit, but I had to confirm later.)[16]

Followed by an impromptu sketch in red pen on notepad of the chemical formulas for oxidation of DOC, dissolved organic carbon, depending on whether the wastewater system used aerobic or anaerobic digestion (symbology a bit over my head).

Next, chit-chat about how "digestion" really does mean digestion, in that, in wastewater treatment, microbes do the work of digesting organic matter, while their human farmers try to grow the best population assemblages for that work. Like all farming, it's nurture as much as nature.

Finally, Professor Bertman rendered this verdict: Infrastructure diagrams were not necessary or even interesting to *him*. Still, he advised, it might be more credible to go a level deeper . . . perhaps identify and explain a couple of the problematic bacteria in pulp and paper production?

Talk about one or two species of bacteria, and he'd be good? What a relief, truly. Thank you, Professor Bertman.

Sad to say but predictable, the relief was premature. Slime-forming species turn out to be the ultimate drama. But before that insight, before slime will exude up its rightful pedestal as an extracellular polymeric substance worthy of awe, a naïf has to flounder for a bit in the microbial worlds of random microbiologists.

For some random floundering, take L. J. Van Dijk, R. Goldsweer, and H. J. Busscher, whose research appeared in the first issue (1988) of a new publication, *Biofouling: The Journal of Bioadhesion and Biofilm Research*. The title of their article was "Interfacial Free Energy as a Driving Force for Pellicle Formation in the Oral Cavity: An *In Vivo* Study in Beagle Dogs."[17] You might read the title again, for clarity (I did)—"Interfacial Free Energy as a Driving Force for Pellicle Formation in the Oral Cavity"—and say in defeat, *Really?* What irony in stalling out at phrases with the driving-force momentum of free energy.

If you're susceptible to detours as well as stalls, you might move past irony and turn post-ironic by wondering, *What about those beagle dogs?* MedicineNet defines *in vivo* as "an experiment done in the body of a living organism." So "an *in vivo* study in beagle dogs" means an experiment done in the body of a beagle. "In fact," say investigative journalists Glenn Greenwald and Leighton Akio Woodhouse in their exposé "Bred to Suffer," "the majority of dogs bred and sold for experimentation are beagles, which are considered ideal because of their docile, human-trusting personality. In other words, the very traits that have made them such loving and loyal companions to humans are the ones that humans exploit to best manipulate them in labs."[18] This was the least hellish, least heartbreaking paragraph in the authors' fierce account of cruelty to beagles and other animals grown in a different sort of "mill" and purposed only for their convenience to the physical infrastructure of research.

Lesson learned: Sometimes you should digress, make yourself vulnerable to other realities, refuse the self-protective armor of irony.

Biofouling and other journals did offer more on-point sources for this amateur's bibliography of biofilm. For instance, in the early 1930s, T. D. Beckwith and J. R. Sanborn separately studied slime-forming bacteria in pulp and paper mills. "Mucilaginous colonies" constituted "a problem of increasing magnitude,"[19] said Beckwith. They were an "invasion of pulp and paper mills," Sanborn agreed, continuing thus: "Once the slime growths have incrusted pipe lines and accumulated in thick, adherent layers in the pulp-handling equipment, chests and tanks, a laborious, time-consuming clean-up must be made, during which the system is rid of these deposits."[20]

Clearly slime was an old problem for experienced companies like Domtar and Dunn. But the problem of biofouling was and still is a messy tangle of information. Journalists untangle informational knots by clarifying the who, what, where, when, and why of a story. Perhaps these Ws applied to biofilm can lead us to the "how" of a cleanup. The "what" of biofilm provides a basic knot to start on.

The what: As filamentous bacteria digest organic matter, they secrete a moist, sticky substance that Beckwith described as "bacterial gum," or slime.[21] The chemical term is extracellular polymeric substance, EPS. EPS consists of polysaccharides, which are long carbohydrate molecules, along with small amounts of proteins, lipids, enzymes, and eDNA. The filaments of the filamentous bacteria provide scaffolding for their collective slime output. Hence these bacteria build their own protective matrix. In a paper mill, the matrix will add wood fibers and other debris.[22]

Filamentous bacteria can build slime quickly. In one article, microbiologist Hans-Curt Flemming and colleagues include a photo of a hand grasping a spatula. While the spatula appears to be scraping batter into a pan, the batter is actually five days of slime from a researcher's "slime board."[23] The amount looks impressive for five days.

A shout-out here to Hans-Curt Flemming himself. Even in super-serious tomes like the fifth volume of the Springer Series on Biofilms, Flemming celebrates the wonder of biofilms.[24] Biofilms are found in 3.5-billion-year-old fossils, he often begins. They are possibly the first and unquestionably one of the "most successful modes of life on Earth."[25] Flemming's intricate "what" of biofilm goes like this: bacteria are single-celled, but with biofilms, don't imagine free-standing cells. Imagine a microbiome, a community of multiple species, with each occupying its ecological niche in the slime.

Flemming argues that biofilms are a continually "emergent life." Does emergent mean they are the embodiment of self-reinvention? Real-life change agents? Yes! Not only do filamentous bacteria form their own habitats within the matrix but they communicate, they cooperate, they have a "bacterial lifestyle."[26] Slimed fishing poles aside, biofilms are extraordinary life-forms.

Extreme weather events are called disasters when they impact people disastrously. Likewise, biofilm is called biofouling when it becomes a problem for industry or human health and activities. Biofouling includes biofilm in a paper mill that stains paper or interferes with machinery; biofilm on a ship's hull that creates drag, raising fuel costs;[27] biofilms dangerous to humans that lurk in food processing facilities, or inside medical devices such as intravenous tubes, catheters, and implants.

Flemming and fellow microbiologists tackle biofouling head-on. Which bacteria are culprits, which are accomplices, which are bystanders? Nearly ninety years of research run from Beckwith and Sanborn to Flemming et al. Yet mill to mill, biofouling presents a new who-done-it mystery every time.

The who of biofouling is a three-fold challenge. First is species identification. Scientists classify bacteria in their laboratories, not in the mill. Such bacteria are "culturable," because scientists can grow them. Slime-formers *Haliscomenobacter hydrossis*, *Sphaerotilus natans*, and *Thiorix* spp. make prominent appearances in studies of culturable filamentous bacteria. In honor of the aforementioned Professor Bertman, here is a longer list:

> *Achromobacter* spp., *Aerobacter* spp., *Acidovorax* sp., *Bacillus cereus*, *Bacillus coagulans*, *Bacillus vesicularis*, *Bacillus vulgatus*, *Bacillus* spp., *Beggiatoa* spp., *Brevundimonas vesicularis*, *Burkholderia cepacian*, *Chlamydobacteriaceae* spp., *Chryseiobcterium* spp., *Cloacibacterium* spp., *Clostridium* sp., *Cytophaga* spp., *Deinococcus geothermalis*, *Enterobacter*, *Escherichia*, *Flavobacterium columnare*, *Flectobacciolu major*, *Haliscomenobacter hydrossis*, *Herpetosiphon* spp., *Klebsiella pneumonia*, *Meiothermus silvanus*, *Methylobacterium* spp., *Microthrix parvicella*, *Nocardia* spp., *Nostocoida limicola*, *Paenibacillus stellifer*, *Pseudomonas viscosa*, *Pseudoxanthomonas taiwanensis*, *Rhodobacter* spp., *Rubellimicrobium thermophilum*, *Sphaerotilus natans*, *Sphingomonas* sp., *Starkeya* sp., *Tepidimonas* spp., *Thiorix* spp.

Other biofilm bacteria are not culturable. Scientists cannot grow these in a lab because they haven't found a medium, or habitat, acceptable to the organisms.[28] Until advanced gene sequencing, unculturable bacteria could not be classified. Many still don't have a place in Linnean taxonomy. In the 1970s, D. H. Eikelboom addressed this problem with a key for filamentous bacteria, classified or not.[29] For unclassified bacteria, Eikelboom type-cast each based on physical characteristics, then assigned a unique four-digit code. Type 021N and Type 0041 are formidable slime-formers, and here are others: Type 0092, Type 0581, Type 0675, Type 0803, Type 0914, Type 0961, Type 1701, Type 1851.

Second, even when suspect species appear in slime episodes, they also inhabit the mill in non-fouling ways.[30] Other conditions might be decisive in biofouling. Surface material is a factor. A few tenacious biofilms all but fuse themselves to certain metal alloys or plastics.[31] Finding repulsive materials with which to make water pipes is no small matter.

Knottier yet, the microbial community (microbiome) might drive biofouling. A case in point: Zumsteg et al. suggest that *Tepidimonas* spp. and *Chryseobacterium* sp. need *Acidovorax* sp. as an auxiliary catalyst for their slime production.[32] With biofouling, fixating on individual species is pointless if cause and effect take place at an ecosystem level. Which leads directly to a third challenge:

We cannot generalize biofoulers across pulp and paper manufacturing.[33] The ecology of biofilm involves more than a web of interacting species. Biofilm ecology includes the totality of how a microbiome interacts with its physical environment. Ergo, the "who" of biofouling is particular to a place, even to a specific mill.[34]

The where: Where filamentous bacteria exist is a different question than where they become biofoulers. The answer to the first is, they're in air and water, and in constant circulation. Mills are open systems. A facility could not hermetically seal itself from microbes any more than we could eliminate the microbial worlds that inhabit our bodies.

On the second question (where they become biofoulers), the answer is, it depends. Within a mill, "white water" is a microbial hotspot. White water is the water drained from paper pulp. For filamentous bacteria, white water is a rich gumbo of organic carbon. As white water cycles through the mill's wastewater treatment plant, or is released as treated effluent into nearby rivers, biofilms attach themselves to all sorts of surfaces. They thrive in places that are out of sight or hard to reach—pipes, ducts, filters, valves, any crevice.

The when: Simple answer: Continuously. (A lesson Domtar and Dunn management relearned the hard way.)

Which brings us to our last W, *why* is biofouling such a challenge? Couldn't Domtar disinfect the mill?

Again, not so simple, which makes this more of a *why not?*

Disinfectants like chlorine can reduce but not rid a mill of biofilm. Flemming sees disinfecting as a medical paradigm for dealing with pathogens in hospitals.[35] But the medical paradigm won't work in a paper mill, he goes on. For one thing, the slime is protective, it's "a fortress."[36] Within their EPS matrix, bacterial communities can respond to threats. They conceal themselves, coordinate to defend themselves, and change behavior when conditions change. They develop resistance, so that an escalator of new biocides must come on line.

What's more, even if you kill a large percent of filamentous bacteria, you leave their dead cells for surviving bacteria to digest. The medical paradigm is to clean, or sanitize. In a paper mill, you're killing but not cleaning. "The term 'disinfection' is thus completely inadequate," Flemming says.[37]

At last we come to a vexing, even perverse, *why* for Domtar: SUSTAINABILITY, of all things.

Strides toward sustainability were facts in pulp and paper production. Recycled paper is now a critical consumer product, easing pressure on forests and landfills. Chlorine-free paper bleaching lessens toxic pollution in waterways. Modern mills recycle their water, which dramatically decreases water consumption. And yet, used paper stock introduces new bacterial contamination. Less chlorine results in more microbial growth. Closed water loops circulate filamentous bacteria through the mill. The paradox of progress was superior conditions for biofouling.[38]

There is "no 'silver bullet'" for biofouling, Flemming stresses.[39] All is not hopeless, but the problem requires a different paradigm. Sometimes Flemming sounds like the Rachel Carson of biofilm. In her 1962 environmental landmark *Silent Spring*, Carson rejected a postwar paradigm that we can control nature with force.[40] Organic phosphates, chlorinated hydrocarbons—Carson saw that such biocidal tools of extirpation will turn against us. We coexist in a web of life, she argued, therefore, to make war on some part of the natural world is to make war on ourselves. But within the web, Carson counseled, we can work with nature more holistically.

The web of life was Carson's alternative paradigm, the wisdom she gifted to a society waking up to the limitations and horrifying consequences of thousands of new manmade chemicals released unchecked, unstudied, and in unfathomable quantities, into our soils, forests, and waters. In journals like *Biofouling*, *Microorganisms*, and *Nature Reviews Microbiology*, Flemming and his peers pass Carson's wisdom forward to more specialized audiences: "Antifouling measures are directed against the oldest, most successful, resistant and ubiquitous form of life on earth. Success is only transient, and we should learn how to live with them."

"Thus," they conclude, "an integrated strategy appears more realistic, successful, and sustainable."[41]

⌒⌒

In "Microbial Biofouling: Unsolved Problems, Insufficient Approaches, and Possible Solutions," Flemming made the kinds of recommendations that Domtar had to adopt as a rear-guard action. An integrated strategy included monitoring ("eyes in the system"), inhibiting ("low-fouling surfaces"), and better mill design ("cleaning-friendly").[42]

The most important part was old-school. As far back as 1933, Sanborn said that "combating slime troubles" begins with "physical means (scraping, hosing out, brushing, sluicing)."[43] Likewise, Flemming advocated for "technical hygiene" and "good housekeeping," both of which meant frequent, laborious, old-fashioned cleaning. Think of it like brushing your teeth or vacuuming your living room, he explained. "There is no way to brush teeth once and forever. If this is acknowledged, it is possible to learn how to live with biofilms and minimize biofouling problems—which requires just some attention."[44]

In Port Huron, some attention was paid. Over Thanksgiving of 2010, when the mill shut down, divers scoured the manifold and discharge pipes from mill to river, then installed a screen at the seawall opening.[45] Visual solids were so dense that early on the company authorized 24/7 monitoring. In addition to divers,

Domtar added the eyes of a video feed. Seeing the system in real time remained a linchpin of Domtar's overhaul.

For better technical hygiene, engineering firm Black and Veatch redesigned Domtar's pipeline system "to allow for frequent cleaning."[46] Domtar also upgraded its pigs and pig launchers. A pig is a mechanical cleaning device. A pig launcher shoots the pig through the pipeline. After still more citizen complaints of filamentous bacteria in its effluent, the company adopted an "intensive pigging schedule."[47]

Domtar tackled its cleanup strenuously but not entirely successfully. Dunn Paper was not exempt either. Local angler groups and the St. Clair BPAC soon turned Dunn's way.[48] Dunn looked slow and reactive by comparison with Domtar's instant response to Gregory A. D.'s underwater footage. In 2016, the Michigan Department of Environmental Quality followed up on new complaints with an unannounced inspection of Dunn Paper. Like Domtar, Dunn had violated its National Pollution Discharge Elimination System permit.[49]

Local fishers became vigilant; they reported filamentous bacteria from either mill.[50] Domtar and Dunn tried new biocides as bacteria built resistance to the old. Commercial Diving and Marine Services kept scouring the discharge pipes. Greg and Kathy kept filming the river.

A Dazzling Discovery

Mission

One day Gregory A. D. got a call from Robert Haas. Haas was a research biologist and station manager for the MDNR's Lake St. Clair Fisheries Research Station.[1] He had heard about their work with the northern riffleshell clam and wanted to know, would they be interested in an unpaid job at Algonac?

In 1996, Haas and fellow station biologist Michael Thomas had begun a study of lake sturgeon in Lake St. Clair and the St. Clair River, where the river shallows and branches into a bird-foot delta before entering the lake.[2] "The goal of our study," they said, "was to obtain basic information about the sturgeon population crucial for protecting and managing this unique resource."[3] Obtaining basic information was more ambitious than it sounds. Research in Michigan had "focused on a few inland locations where small spawning groups of sturgeon [were] readily accessible in shallow rivers."[4] Whereas the St. Clair Station worked in a complex waterway of lakes, rivers, and delta.

Scale mattered. The St. Clair–Detroit River system was the connecting channel for the longest open waterway in the Great Lakes; from Lake Erie to Lake Huron, free-ranging fish passed unimpeded by dams or other barriers.[5] Moreover, the St. Clair River delta was the largest freshwater delta in North America, which made it a hotspot of biodiversity. That's a lot of "longest" and "largest." At this scale, information on lake sturgeon was shallow and bisected.[6] Two generalities formed the only system-wide knowledge. One, lake sturgeon had survived two centuries that nearly wiped them out. And two, while a remnant population remained, the fish was threatened or endangered throughout its range.[7] Fisheries scientists knew little else: current abundance, spatial distribution, spawning sites, genetic groupings, migratory patterns, fish behavior, or life histories.[8]

During the late nineteenth century, the Great Lakes bore a boom and bust of sturgeon extraction every bit as severe as the white pine boom-then-bust also underway. By the 1890s, Great Lakes sturgeon populations had crashed. Lake Huron fisherman Alexander Truedell remembered when there were "schools of sturgeon in the Pigeon river so thick that their fins were above the water, but that is ancient history."[9] Lake Erie told the same story. From five million pounds in 1885, the sturgeon catch plummeted 80 percent by 1895.[10] Meanwhile, the liquidation of both lake sturgeon and virgin white pine swept west to new waters and forests in Wisconsin, Minnesota, western Ontario, and Manitoba.[11]

Before 1860, commercial fisheries saw lake sturgeon as useless bycatch. Sturgeon were scorned, said ichthyologists W. J. K. Harkness and J. R. Dymond. "Not only were they worthless, but the weight of the big ones and the bony, serrated plates of smaller ones injured the nets so that it sometimes required days to repair them after a catch of several sturgeon."[12] One story of the era passed from grandmother to grandson (who became a fish hatchery superintendent) to Harkness. "His grandmother, who was born about 1850, used to live at Point Pelee," began Harkness. "She remembered when sturgeon came to the bar off the Point in May and June in such numbers that her father, standing in a flat-bottomed boat, killed numbers of them by hitting them on the head with an axe. Only the largest were taken."[13]

Even as late as 1872, a Michigan fisheries agent named James W. Milner observed this mindset in Green Bay, Wisconsin. "A few fishermen," Milner said, "are considerate enough to lower the corner of a net and allow [sturgeon] to escape, but the commoner way is to draw them out of the net with a gaff-hook and let them go wounded, or to take them ashore and throw them on the refuse-heap, asserting that there will be so many less to trouble them in the future."[14]

In 1974, Michigan fisheries chief Wayne Tody surveyed the historical carnage—discarded lake sturgeon stacked ashore "like cordwood . . . set afire and burned, much in the manner that settlers on southern farms disposed or tree stumps in clearing their lands . . . [or] taken aboard boats and burned in the boilers. And, of course, there are always the stories of attempts to dig them into the ground for use as fertilizer."[15]

"Today, we deplore the slaughter of the passenger pigeon, the American bison, and other species of our wildlife heritage," stated Tody. "But, very likely, no single animal was ever subjected to such deliberate wanton destruction as was the lake sturgeon."[16] While lake sturgeon would eventually become a prized fish, he decried, "it is pathetically unfortunate that sturgeon markets were not discovered until the great standing crops of this fish had been destroyed."[17]

Sandusky, Ohio, on Lake Erie, was the first to establish a major commercial sturgeon fishery in the Great Lakes.[18] Again from James Milner in 1872, but this time on his visit to Sandusky (which made the Green Bay fishery seem backward): "There were about fourteen thousand mature sturgeons handled, weighing about seven hundred thousand pounds, obtained from about eighty-five pound-nets."[19] Sandusky reported 720,500 pounds of lake sturgeon that year.[20] Chicago was the only other place to list a sturgeon catch, a fraction of Sandusky's at 25,147 pounds. Thereafter, fish dealers couldn't keep up with demand.[21] "From a despised fish fit only for the offal pile, the sturgeon has become of great commercial value, and in certain localities is fished for exclusively." So reported Michigan's State Board of Fish Commissioners in 1890.[22]

The sturgeon body, deconstructed, became a marvel of the marketplace: sturgeon meat smoked as a delicacy, sturgeon skin tanned for leather, sturgeon oil substituted for sperm whale oil. "Nearly every part of [the sturgeon] is utilized in some way," the commissioners noted. "The Chinese prize even the dorsal chord, which is cut in slices and dried and used as food."

More lucrative than dorsal chord was the sturgeon air bladder, also called the swim bladder, because it allows sturgeon and other bony fish to adjust their buoyancy as needed. Scuba divers like Greg and Kathy use a buoyancy compensator with an inflatable bladder to do the same thing. Inflate the bladder to rise, deflate the bladder to sink. A membrane in sturgeon air bladders contains a special collagen called isinglass.[23]

In a talk on isinglass in 1905, Professor T. W. Bridge said that, "the air-bladder, or, as it is sometimes termed, the 'sound' or 'float,' is in some respects one of the most remarkable of the internal organs of fish. Essentially, the air-bladder is a sac with membranous walls and gaseous contents, which, in its natural position in the body, is situated above the stomach and beneath the backbone."[24] Bridge's talk was titled *The Natural History of Isinglass*. He delivered it in Birmingham, England, at a meeting of the Institute of Brewing.[25]

Isinglass could be made into glue, plaster, or pottery cement. Isinglass thickened food (it was the gelatin of the day). And, to explain Bridge's audience of British brewery owners, isinglass quickened the settling of sediment during beer and wine production. "The great importance of isinglass to the brewing industry is due to its value as a clarifying agent in the manufacture of beer," said Bridge, "and apparently it is the only product of the animal body which can be used for this purpose."[26]

Preparation was simple, he told the audience.

The air-bladder is cut out of the fish and washed, the outer or peritoneal membrane is stripped off, and the bladder is then dried while partially distended; or the bladder is slit open lengthwise, flattened out and pressed, and afterwards dried as "leaf" or "book" isinglass. In certain parts of Russia the air-bladders are prepared by steeping them in water, removing carefully the outer coat and any blood that may be present, then placing them in hempen bags, squeezing and softening them between the hands, and finally twisting them into small cylinders. After being dried in the sun, and bleached by the fumes of burning sulphur, they are ready for the market.[27]

The isinglass market intensified commercial sturgeon fishing. But outright mania ensued with a caviar rush, which turned sturgeon eggs into black gold. Caviar, commented Michigan's fish commissioners, was "a dish formerly prized chiefly by the Russians and their neighbors, but now often finding a place on the menu of the best hotels and restaurants." Whereas a keg of caviar in 1885 cost nine to twelve dollars, the same keg at the turn of the century cost over one hundred dollars.[28]

In his talk on isinglass, T. W. Bridge had seemed fascinated with the transformation of a sturgeon air bladder into this most versatile of products. So too did the commissioners seem fascinated with the transformation of sturgeon eggs into this most elite of international gastronomy. "The eggs are rubbed with the hand upon and through a course sieve, to separate them from the connecting membrane, and then a fine german salt is added, the proportion being a trade secret, and the product thoroughly stirred with the hand and drained. It is then dried and packed in kegs for shipment." One such shipper would have been Neilson and Company, which had a caviar factory in Algonac, on the St. Clair River.

Nor did the commissioners neglect where and how lake sturgeon were extracted:

> The sturgeon is taken principally in pound nets, but many are caught also in the narrow and deep channels of the St. Clair Flats, and elsewhere, by set lines. A strong line is stretched upon stakes driven in either bank of these channels, or anchored at the bottom on either side, and from this depend many shorter lines to which are attached large hooks which rest on the bottom. The sturgeon in rolling upon the bottom becomes entangled in these hooks and is captured with a gaff.

A gaff is a long pole with either a spear or big hook at the end.

The result was both gruesome and unsustainable. Margaret Beattie Bogue cites marine biologist Henry Frank Moore, who in 1894 "lashed out at the 'barbarous' methods of grappling for sturgeon in these locations. They mutilated the fish, tearing away large patches of skin and flesh and ripping out eyes."[29] Indeed lake sturgeon "were taken by every available means," said Scott and Crossman, "from spearing and jigging to set lines of baited or unbaited hooks laid on the bottom, to trapnets, poundnets, and gillnets."[30]

Circa 1880 to 1900, Great Lakes sturgeon catches plunged from 7,841,000 pounds to 1,772,000 pounds. Michigan sturgeon catches: 4,300,000 pounds to 140,000 pounds. In the Huron–Erie Corridor: 1,091,000 pounds to 92,000 pounds.[31] Since the numerical unit was pounds of sturgeon handled, pounds were the proxy for what was happening to numbers of living lake sturgeon. In those hungry decades, the fish themselves weren't counted; they were countless.[32] They were just there, for the trashing and taking, because there were so many. Today we estimate that lake sturgeon in the Great Lakes plunged from fifteen million to fewer than 150,000 fish, or down 99 percent.[33]

In 1973, the Michigan state legislature passed a resolution "commemorating the 100th anniversary of the Department of Natural Resources Fisheries Division."[34] The resolution began, "Whereas, a century ago, on April 19, 1873, the State's first fish commission was appointed. The same year the first fish hatchery in the State of Michigan was built at Pokagon and named Crystal Springs for the clear water supply found there." The resolution charted a century of progress in fish management—"fish culture as a science," "modern fish hatchery facilities," "scientific management techniques." Yet buried within the resolution was an apologia:

> Whereas, the general concept of our fishery resources was one of inexhaustibility in the late 1800s, the early fish commission recognized the perilous position of the Great Lakes food fishes and the Michigan grayling, and made many valiant attempts to save these resources from becoming extinct.

The resolution named the Michigan grayling but didn't quite come out and say that this most gorgeous of freshwater fishes went extinct in its native Michigan waters. Today, Michigan grayling is more correctly known as the Arctic grayling. A consortium hopes to redress grayling history and reintroduce Arctic grayling to the Big Manistee River. Success is uncertain and could take decades. After all, the grayling of long ago knew a different Manistee. Let's hope the river welcomes them home.

Lake sturgeon didn't suffer the Arctic grayling's fate, those futile reports of last sightings in the wild before the fish finally froze into fiberglass replicas mounted on the walls of fish hatchery visitors centers. But during the twentieth century, new threats kept sturgeon on the brink: spawning grounds destroyed from dredging to expand international shipping; food-webs disrupted from introduced species; reproduction impaired from industrial pollution; and covert poaching in places like Algonac.[35] Combine those with a vulnerability of sturgeon biology: slow maturity into reproductive adulthood. Grieved Wayne Tody in 1974, "It is a sorrowful commentary that the sturgeon could exist and thrive in North America for 50 million years before tangling with the greatest of all predators—man."[36]

And yet, miraculously, it's still here. The species persisted. The question was, to what extent?

———

For their pilot assessment in 1996, Thomas and Haas got leads from anglers and divers. Where the delta opened into Lake St. Clair, there was a "sturgeon hole." Thomas also called the area Sturgeon Central.[37] They trawled for sturgeon, dragging nets behind the *R/V Channel Cat* (R/V means research vessel). Fortunately, they saw "unexpected success" early on.[38] Unfortunately, they didn't have a holding tank onboard for the fish. The team improvised. A photo shows a line of shallow rectangular tubs on deck, a water hose filling a tub, a bucket and a scale for weighing off to the side, and pairs of sturgeons draped across the edges of two touching tubs, tails dangling in the first tub, heads underwater in the second.[39] After that, the *Channel Cat* had a tank.

May to June 1997, the scientists again caught lake sturgeon in the sturgeon hole.[40] They measured and weighed each fish and noted any sea lamprey scars, tagged the fish with a monel cattle ear tag (for a unique identifier if the fish was recaptured), and took a pectoral fin ray to establish age.[41]

Another promising area was the North Channel of the St. Clair River, by Algonac, where sturgeon would porpoise in early spring. By contrast with Lake St. Clair, the St. Clair River had strong currents and steep bottomlands. Like fishers a century before, Haas and Thomas decided against unwieldy nets. Instead they used baited setlines for their survey (six lines, twenty-five hooks per line). The weekly schedule went like this: On Monday between 9:00 a.m. and 4:00 p.m., string the setlines cross-current across the bottom. Aim for "the widest scent plume downstream."[42] Tuesday through Thursday, pull the previous day's setlines, record data on caught sturgeon, rebait empty hooks, put setlines back in. Friday, pull the setlines for the weekend.

According to Thomas and Haas, most sturgeon were "hooked in the mouth" and released without injury or only "minor damage to skin and muscle tissue at the point of hook penetration." Another 20 percent suffered deeper wounds, because they "were foul-hooked or snagged . . . in the belly, side, or even the tail." Foul hooking peaked during spawning, when the fish writhed and rolled on the bottom as they mated. The biologists did note that "all lake sturgeon collected were released alive."[43]

The researchers also experimented with baits. Relisted here from "Capture of Lake Sturgeon with Setlines in the St. Clair River, Michigan"[44]: Cut baits: bluegill, Atlantic herring, common carp, white sucker, channel catfish, northern pike, and alewife. Whole fish baits: alewife, rainbow smelt, spottail shiners, trout-perch, and round goby. Miscellaneous baits: chicken livers, earthworms. Fourteen different baits—a sturgeon deli, and a cliffhanger of a question. Which would they go for? Out of eighty-four lake sturgeon, one chose the cut channel fish. As for the remaining eighty-three? They opted for that recent arrival from Eurasia, the round goby.

For their 1997 field season, Haas and Thomas received a grant from the Federal Aid for Sport Fish Restoration Program. One objective was to "characterize spawning habitat in the St. Clair River." But first they had to find spawning grounds.[45] Anglers pointed out a favorite part of the river for sturgeon fishing at Algonac. The spot was near shore, facing three small close-set houses. The Maslanka family gave Haas and Thomas permission to use the dock in front of their house.[46] In that humble place, the team struck their own black gold when they pulled spawning sturgeon into the *Channel Cat*. "Eggs all over the boat, milt all over the boat," Haas told Greg and Kathy. For proof of site, the scientists needed fertilized sturgeon eggs from the river bottom. That's where Gregory A. D. came in. Can you find eggs? Haas asked.

"We had no idea what size a lake sturgeon egg was," recalls Kathy.

Oh it's caviar, the researchers told them.

Greg and Kathy grumbled that "we're not in the caviar crowd." Still, they assumed a *we totally got this* attitude and jumped in. In no time the duo brought up a big coal clinker inlaid with eggs. The researchers "spent a good fifteen minutes telling us how great we were," says Kathy. But then, "Mike turned to us and said these were sucker eggs" (redhorse suckers, and much laughter from Greg and Kathy).

"We need black or gray, same size, you're good on the size," Thomas instructed.

"We went right back in and brought them up."

Haas and Thomas wanted to know, how many lake sturgeon used this surprise of a coal-cinder reef? What were its characteristics? Could it be replicated elsewhere for restoration efforts? They got funding for Kathy and Greg to install a camera. The two were excited about the mission. They had seen lots of lake sturgeon but not a sturgeon spawn.

With ten feet of loose cinders, the reef was a hard site to maneuver around. Greg and Kathy spiked a rake handle for hand-holds while they worked. They used a metal box to protect the camera from sturgeon jostles and installed it at the edge of the reef.[47] One worry, said Jerrine Nichols, a scientist for the United States Geological Survey (USGS), "was that the sturgeon would react negatively to the presence of the SCUBA divers." The opposite happened. "In fact, spawning sturgeon were greatly attracted to the divers and would rub up against them, allow themselves to be scratched and in general treated the divers as another large sturgeon."[48]

Greg and Kathy got eighty hours of footage, the full spawn, "sturgeon laying in front of the camera, resting in front of it, spawning in front of it." The team estimated thirty to fifty spawning sturgeon at the reef.[49]

Discovery

Ecological research in the Great Lakes is a power train of state and provincial agencies, federal agencies, and academia. Over time, the USGS Great Lakes Science Center appended a telemetry study of lake sturgeon movements to the MDNR's Lake St. Clair Fisheries Research Station assessment.[50] One graduate student from the University of Michigan carried out research for the telemetry study, and later joined the US Fish and Wildlife Service (USFWS). University of Michigan, USGS, and USFWS scientists built on lake sturgeon research with predation studies of the round goby in the St. Clair and Detroit Rivers. NOAA's Michigan Sea Grant called all hands on deck for a major restoration initiative to install artificial reefs in the two rivers.

Without a doubt, Kathy and Greg formed opinions as they connected with this research engine. To them, the DNR guys were "real": they were "on the water . . .

doing real research . . . handling fish." The Feds, USFWS, and USGS, they were under pressure to be the experts, to win huge federal grants for restoration projects like artificial reefs (that were unlikely to succeed in Kathy and Greg's view). And university types? "We got no use for academics, Lynne, I'm sorry." Except, that is, for David Jude, a fishery biologist from the University of Michigan. They both nod. "Jude was great. He realized that we [brought] something to the table, that we could be his eyes."

In 1990, anglers and Jude identified non-native round and tubenose gobies in the St. Clair River.[51] For thirty years, Jude would study their impact on Great Lakes' ecosystems. Separately, Greg and Kathy had been trying to get a shot of the river bottom without gobies. The previous year, they explain, "there were so many gobies we were pushing them aside." This was a nuisance because they had a book coming out, and didn't want gobies in the cover photo. Early on, Jude hired Gregory A. D. to be his eyes underwater for preliminary goby assessments in the St. Clair and Detroit Rivers.[52]

Before planning their Detroit dives for Jude, Greg and Kathy wanted to observe the assessment process. So the two accompanied Jude's team to the Detroit River, where they boarded the Environmental Protection Agency's R/V *Lake Guardian*. They thought the *Lake Guardian* was a most impressive vessel, practically a mini freighter, with science labs and even derricks. The plan was to get a preliminary goby count, recalls Kathy, but when the researchers pulled the nets, "they didn't get a single frickin goby." *Were round gobies even in this location?* the group wondered. *Maybe not.*

Greg and Kathy thought otherwise. Here's what they pictured as everyone pondered no gobies: a contraption of collection nets weighted with a wide, heavy board. Dragged along the bottom it would be like a bulldozer, really loud, making the "krruuuuuuuh" of crunched gravel. The gobies "don't know what it is, but all of a sudden coming upstream at 'em is something going 'krrruuuuuuuh.'"

Kathy, impatiently, "He's gonna get the hell out of the way, every fish is gonna get out of the way. . . . Greg just finally suited up, got the camera, jumped in, filmed right where they were saying there was no gobies, and there's gobies everywhere."

Loud sighs from both. "They don't believe us half the time. We're not biologists, so what do we know."

———

Short surges of exasperation: That's not so uncommon between locals with particularized knowledges and researchers advancing generalizable knowledge. The confluence of scientific knowledge with local or traditional knowledge can be turbulent.

At the extreme are structural imbalances and vulnerabilities that have, historically, enabled intellectual and economic plunder, or biopiracy. Biopiracy occurs when researchers extract traditional knowledge of plants and animals from a community, a cultural group, a country, without crediting that foundational contribution to the science, and without sharing the economic benefits—intellectual property rights, corporate gene patents, new drugs, modified crops. In the context of Indigenous or traditional ecological knowledge, philosopher of science and poet Laurelyn Whitt calls this "extractive biocolonialism."[53]

Less extreme but sometimes destructive: Community members and scientists may experience rough relationships during environmental crises. Or scientists themselves may clash. The recent lead poisoning disaster in Flint, Michigan, is an example of both. In 2015, at the behest of Flint resident LeeAnne Walters and others, Virginia Tech engineer Marc Edwards launched an unfunded, volunteer-driven "Flint Water Study." The Virginia Tech team trained community members in emergency citizen science. Their research exposed dangerous lead levels in Flint's municipal water system. Edwards's work with Walters became a model of scientist and citizen who took each other seriously.[54]

A disclosure: Soon after Michigan learned the magnitude of water poisoning in Flint, a group of us helped bring Edwards's doctoral assistant Siddhartha Roy to Western Michigan University. Roy, an expert on copper in aging water infrastructures, had put his own PhD research on hold to coordinate the Flint Water Study. He told a shocking story of official negligence as lead from old pipes leached into the water of Flint homes. But the story also epitomized an ideal: hard-nosed science welded to values like justice, truth-seeking, and community well-being. Sid Roy the scientist was not afraid to stand up in front of three hundred people and also be Sid Roy the passionate human.

That's why I was taken aback two years later when, at an environmental justice summit in Flint, I learned that the relationship between Marc Edwards and some of the Flint community was shattered. Edwards was estranged from many Flint water activists, and also from Michigan researchers, who, following the crisis, began a new Flint Area Community Health and Environment Partnership. The conference room fell tense at even the allusion to him.

Reporters had already noted a rift some eighteen months before the summit, when actor Mark Ruffalo's nonprofit Water Defense spread false claims about *other* dangerous chemicals in Flint tap water. Edwards denounced Ruffalo as an "A-List Actor But F-List Scientist."[55] "Mark Ruffalo should be ashamed of his unscientific fear-mongering," he said.[56]

Sounded reasonable.

But where Edwards saw junk science, enemies of good science, and profiteers of science, locals saw a scientist (Edwards) attack competitors, an egoist claim the limelight, and an outsider ignore their own centrality and self-determination. Self-determination included participating in the community research partnership; Edwards's disapproval was irrelevant, except for his lawsuits and barrage of Freedom of Information Act (FOIA) requests related to the credentials of the lead researcher. "From Hero to Pariah, Flint Water Expert Fights for His Reputation," headlined a 2019 *Detroit News* article about Edwards that practically gaped at the hydra-headed feud: e.g., "The dispute involves lawsuits, open letters, death threats, alleged spurned love and accusations of plagiarism and falsified water data. It has played out in emails, blogs, social media, science conferences and scholarly journals."[57]

While Flint was exceptionally volatile, some acrimony would not surprise any scientist or citizenry immersed in their own water controversy.[58] Still, Kathy and Greg's exasperated "we're not biologists so what do we know" was different. It spoke to a subtle tension, not a stochastic explosion. This was a soft struggle between professional ecological knowledge and the local knowledge of practical experience.[59]

Biodiversity educator Elin Kelsey analyzes this third uneasy relationship. Kelsey identifies a "science first" hierarchy, in which scientific interpretation is elevated while "other types of knowledge held by the public, such as traditional or lay knowledges, are undervalued."[60] Kelsey views this as a problem for biodiversity efforts. She argues that "the complexity and controversial nature of biodiversity issues," demands more participatory, less hierarchical models of information sharing and knowledge production.[61]

In the St. Clair River, a short struggle between scientific and experiential knowledge would reach a breathtaking climax. Greg and Kathy, and the scientists within their orbit, would change what we know about lake sturgeon in the Great Lakes. But first, we have to go back to just after Gregory A. D. filmed the cinder reef at Algonac.

—

Thomas and Haas's study had enriched Greg and Kathy's decades of underwater experience. Afterward they reflected on what they knew of lake sturgeon in the river. You could see sturgeon year-round in two places, Algonac and Port Huron, under the Blue Water Bridge. Logically, there *could* be a spawning site near Port Huron, yet why had they never seen a spawn? The issue might be water temperature, they reasoned (if there was a site). They now knew that lake sturgeon spawned when water temperature reaches approximately fifty-five degrees or higher. Greg's first dives of the season were early, but maybe not early enough? Maybe he went in after the sturgeon finished spawning.

The following spring, Gregory A. D. tried out their hypothesis. Greg dove when the water temperature hit fifty degrees, still too cold for spawning. They checked every day while the water gradually warmed, until, at fifty-three degrees, "OMG!" Kathy exclaims.

"*Thousands* all of a sudden just arrived, and it was like nothing we had ever seen in our entire lives, and we couldn't believe it."

"And the biologists didn't believe us," she adds.

They called Jim Boase, one of the scientists they worked with in the St. Clair. He did not believe them. *No, really,* they said, *thousands and thousands of sturgeon.*

They showed him film footage. *Whoa!* Boase called other sturgeon scientists. They did not believe *him.*

In the natural sciences, consensus emerges incrementally, over decades or longer. Modern ichthyology is closing in on two hundred years of cumulative research. For modern ecology, more than one hundred years have passed since ecological pioneer Henry Chandler Cowles began his famous studies of plant succession in the Indiana Dunes of Lake Michigan. Many generations of working scientists have helped the rest of us understand the natural world in all its thrilling complexity.

Sometimes, though, consensus becomes boiler plate, something everyone just *knows.* And sometimes, what everyone *knows* isn't quite right. Fresh eyes, different angles, other kinds of experience and expertise—Greg and Kathy brought these to the current state of sturgeon knowledge.

The problem was water depth. Until Gregory A. D.'s discovery, the assumption was that sturgeon spawned in shallower waters. By contrast, Greg and Kathy went looking sixty-feet deep. The site did have fast current, and for lake sturgeon, fast current was the precondition, not depth. But those insights came later.[62]

In the moment, everyone beheld a priceless treasure. Each spring, until now unseen and unknown, the largest and most important sturgeon spawning site in the Great Lakes vibrated with life. At least 29,000 lake sturgeon. An ecological Atlantis.

Currents

Rivulets

Greg: *My brother-in-law, he could tell I was kind of a strange kid . . .*[1]

I really didn't talk. Well I never talked in school, and I always felt I was a little autistic or I have some kind of flaw somehow.

But he brought a tank over when I was thirteen, probably to a pool, to a friend's house, and showed me how to use it, and I totally fell in love with that.

I'm not sure if the camera came first, but it was right around the same time he gave me his old little camera. And then once I got into scuba class that instructor gave me a little camera housing that he never used.

I had to lie to get into scuba class cause you had to be sixteen and I wasn't yet, so we kept calling the instructor, who became my best friend and mentor. He knew I was too young, but he let me come in."

Kathy: Your dad lied. Your dad said he couldn't take it anymore.

Greg: The combination of it all, it pretty much cemented where I was going for the rest of my life.

Outlaw

Greg: *That was a good time to quit . . .*

Back in the day I gotta admit I did spear some walleye illegally, but we all did. I mean that was just what you did back then. And then all the sudden they changed the rules and they could take your vehicle and your scuba gear and all that. So that was a good time to quit spearing fish.

Wet-Suited

Kathy: *You love your wetsuit . . .*

Greg: I just love a wetsuit. You're warm when you're underwater and you're just one with the water. That's what's really nice about it.

It feels like a second skin. It definitely keeps you nice and warm. With the current I tend to strap my gear on rather tightly just because as you go deeper the suit compresses and your weight belt loosens up and everything loosens up. I like to make it really super tight.

I'm one of the divers who when you come up after a dive I like to stay in my suit—some divers like to get out and that's the first thing they do—I like to stay in my suit cause it's warm and to me it's a just second skin, and I've just spent so much time in a wetsuit that to me they're not uncomfortable.

Kathy: You go streamlined and you go heavy, too.

Greg: I tend to go heavy. It depends what you're doing. If you just wanna drift with the current, you put enough weight on that you're heavy, but all's you have to do is add a little air to your bc [buoyancy compensator] and then you can become buoyant, you can become neutral, whatever you prefer at the time, whatever you're doing. If I'm looking for stuff then you wanna be heavy. But if I'm trying to move with the fish, the sturgeon or something, you wanna be a little more light, little more buoyant, so you can keep up with them. Especially when you've got a big camera, you've gotta try and move with them. So you wanna be as close to a fish as you possibly can.

Underwater

Greg: *When I'm underwater . . .*

I feel like a superhero, cause I can leap tall buildings in a single bound, leap over wrecks.

You can fly, you can float, and hover.

It's amazing because you can run, you can bicycle, you can do all kinds of things underwater. You're kind of like Iron Man where you can take off and just . . . fly.

It's quite amazing.

Current

Greg: *The current makes all the difference in the world . . .*

Kathy: When I first started diving the river what Greg told me was: "You can never get lost, you can always come home. Because all you ever have to do is stop. If you're confused and you lose track of where you are and which way is which, you stop and fan the current, and even if you're downstream where you're in a really low current area, if you fan soft sediment it will go with the flow. Wherever it's flowing put your head up, go to the left, you're coming home. Go to the right you're gonna go to Canada. But come to the left you're coming home."

Hands

Kathy: *You want to work with the river, never against the river . . .*

Greg: You try not to swim with your hands. That shows a novice diver if the hands are moving much. I'm pulling myself most of the time. I usually only have one hand free, so everything's one-handed.

Even if you're drifting from a boat or drifting down the river alone, and you're with the sturgeon, the current gets you moving so fast that you try and slow down. I drag my fins flat against the bottom, and then a lot of times will put a flat rock in my one hand, and I force that into the gravel. And I'm going head first downstream, feet are dragging, and I've got my hand on a rock, and it's plowing a trough as I try and slow down. Then I'll be filming the sturgeon, and sometimes I'll turn the camera back on me. But you still get going so fast that if you try to maneuver around to the sturgeon—they're just so quick and free flowing—I want to be as slow as I can.

Kathy: You use the current a lot. You want to work with the river, never against the river. As long as you're working with her, you can do some amazing things. But the second you try to work against her, she will beat you.

Feet

Greg: *I consider myself with webbed feet . . .*

I wear real long large heavy fins.

You would think, oh you're kicking down there in the current all the time. Actually I hardly ever kick. I would say 80 percent of the time I'm not kicking. That's because I want to stay as long as I can, and if you use your big powerful muscles you're gonna consume a lot more of your air out of your tank.

So I find that my feet are very content just to lay my fins right down on the bottom. It keeps me centered, and it keeps me from moving too fast out of an area. If I want to go upstream I might kick just a little to get to a hand-hold and then it's all pulling. Very seldom do you really kick, unless you're going up the wall or at the surface. Very little kicking is my technique.

Kathy: People who aren't familiar with current burn themselves up. Each stroke is like running. If you kick for five minutes straight you start breathing hard. You're burning up your air and you have a limited supply.

You [Greg] use your legs. You jam your knees in and you wedge your legs and you wrap your legs around stuff. You're dug in.

Greg: Yeah, it kinda depends what you're doing, but filming or looking for rocks in the current, where you gotta stay there and move sideways in the real hard current. . . . The tops of your feet are pressed totally hard onto the ground cause you're on your stomach. Your feet are hard on the ground, your elbows are trying to catch any rock they can to stay. It's not uncommon to try and wedge yourself under something.

And if you're trying to go upstream to get that one perfect rock that's just out of your reach, you can't sit there and kick cause you'll just freak yourself out. And next thing you'll be popping to the surface cause you're either out of air or exhausted.

So you gotta work your way up. It's that rock climbing in a waterfall idea. When you do use your hands, if you grab a rock, you don't just grab it and pull yourself. You push down on the rock. Otherwise the rock will come with you if you pull against it.

There's a lot of techniques underwater. Ferry gliding across the current like fish do, you learn to tilt your body and your shoulders, one up and one down. That'll ferry you across current where you don't even have to kick. There's all kinds of little tricks that become so natural.

You're really aware of the fins. This last sturgeon trip I was using fins that I usually use for dry-suit diving. In a dry suit you don't really go anywhere in the river cause it's just a pain in the butt to do stuff in a dry suit in a current. But in a wetsuit I was wearing the wrong fins, and they have slits in 'em. Supposedly that gives you less resistance, but you can also get caught up on wreckage—the spikes tend to end up right in there—and then you're really aware of your feet cause, in the current if you're stuck on something and there's a rod through your fin, and your head's downstream, it's very difficult to get your foot out of that position. And that happened to me on the last sturgeon dive. And-uhh, that wasn't good [small laugh], that wasn't good at all.

Kathy: No.

No, his explanation when he came up from that dive was, "I almost died." That's the first thing he said to me. I hate it when he says that, cause I know, he ain't kidding. He's only said it about ten times in thirty years.

And he was like, "yeah I got pinned." So you imagine, all that force is against your body. It's like hanging off a skyscraper. You're by your foot, and it's pulling, so he's now stretching like spaghetti man and he can't get the fin off.

Boom, it reached the pressure point.

Greg: Everything popped at once.

Lynne: How did that happen on a sturgeon dive?

Kathy: Oh, sturgeon like the shipwrecks. That's one of their spawning sites. The *M. E. Tremble*.[2]

Greg: No, this was the *Fontana*.

Kathy: Ohh, you were over on the *Fontana*?? Ohhh, well shhi, that's even worse. Okay that's on the Canadian side, that's like right under the, that's the hardest, that's the hardest wreck to, oh yeah, you're done.

Greg: Yeah I lost my fin. That's the first time I ever lost a fin. I continued the dive. I had to swim up with one foot. Then you really realize how important a fin is when you only got one.

Breath

Greg: *It's called skip breathing . . .*

I am aware of my breath, because I'm gonna try and get to a totally calm state where I can make the air last in my tank as long as possible.

Kathy: What it really is is yoga breathing. Half of yoga is about breath control. So it's being conscious of the pace of your breathing, the depth of your breathing.

We're also very aware of our bubbles, our exhalations, because of the fish. So when we're filming that's critical.

Most fish don't like bubbles. I imagine to them it's like little grenades going off.

Greg's trying to film a beautiful school of minnows passing over his head between the sunbeams and him, he's literally gonna hold his breath, even though this goes completely against all of our dive training. Everything in diving says you never hold your breath. But the reality is, if you're lying flat on bottom, in that situation, every shooter in the world holds their breath. The second you exhale the school splits.

Greg: I will just, *ssp, ssp.* Like those gizzard shad, where they're just thick. Well the minute you exhale—*psshew*—they're all goin away. So you try to film as long as you can. Then you need a good big breath.

Kathy: An average diver trying to do skip breathing or trying to hold their breath, hugely dangerous, probably going to kill themselves.

If you're out on the sturgeon dives and you're doing that, doing controlled breathing, and not breathing heavy and stuff, and he's done this, and then all of a sudden he runs out of air because he's so into filming and he's not even paying attention . . .

Greg: That happened.

Kathy: And that has happened. Then he would have to have the skills and the training and the mental capacity to remain calm and make an emergency ascent from sixty foot. That can and has killed a lot of divers.

Greg: Last ten feet is tough.

Kathy: Right. It's tough. Cause everything in your brain is saying *"breathe, breathe, breathe, get to the surface, I mean you're out of air."* When your lungs are empty, your brain wants nothing more than to breathe. And so it wants to get you to air as fast as possible. And that's what will kill you. You have to go slowly to the surface even though you're out of air. You have to completely go against your flight instinct, and you have to remain calm so that you can do a controlled ascent. That last ten feet, now you can see it, it looks like it's right there, and you wanna just, "uhhh," and if you do you'll probably kill yourself.

So the whole thing can be very risky. Those are some of the things that we try very hard to avoid. But, if you're gonna do this and you're gonna put yourself in those situations, you better be able to handle the consequence. It's like don't do the crime if you can't do the time. That's how Greg dives, like he's prepared to deal with the emergency. He's not unaware that he's putting himself in high risk situations. Which is also why he doesn't take anyone with him.

That's part of why we don't have children too. We're not gonna have a kid and then potentially let their dad die. We're very honest and open about that.

Greg: I never thought I would live this long, because of the diving. Between carpentry, ten stories up and scared to death, and then all the diving, some sailboats trips. Sixty, are you kidding? I'll never make it to sixty. Here I am at sixty-two. Wishin I didn't hurt so much. All my joints and two back operations and herniated disks. Wish I would've taken it a little easier, but here I am.

Freighter

Greg: *A shadow from a freighter can be very disconcerting . . .*

Kathy: You hear 'em coming a long way away. They're very loud. They're very, very loud. It's like a siren. You hear it coming waay before you see it. When it gets loud enough you start thinking, "oh where's he coming from?" So it's the same thing for us. First you hear that very faint thump, thump, thump. It's way off in the distance. "Oh, freighter's coming." You don't know if he's upbound or downbound, you just know somebody's coming. When it gets to a certain crescendo, you still don't know which way he's coming, but you reevaluate where you are in the river and what you need to do. If you're in St. Clair, or even Marysville or anywhere from Marysville downriver, as soon as you hear the light thump, you should be thinking about heading in.

It's not uncommon to have two or three go by on a single dive. We always say, aww it's no big deal, because they draw thirty foot when they're fully loaded. Most of the time we're diving in fifty to sixty foot. So you've got twenty to thirty foot. I say that to people thinking that's a lot, and usually most people look at me and they're horrified. That's two stories, though. If you looked up at a two-story building, you wouldn't necessarily worry about the antenna falling on ya.

Greg: Thump, thump, thump. Sometimes it's extremely loud. But then again, even a thousand-foot freighter coming downbound under the bridge can sneak right up on you. You wouldn't think something that big could do that.

Kathy: Downbound they tend to turn their blades off because the current's so fast that they don't need propulsion. Coming down Lake Huron entering the St. Clair River at the mouth, they get goin and the current does grab 'em and it's very narrow and they have to make a sharp bend. They're just using their thrusters to control their direction and letting the current carry 'em. One of them definitely caught Greg once. That was the one time you really thought you were gonna bite it.

Greg: I was way, way, way out, and I was diving from the American shore, and there's this one place that I found all these great Indian net sinkers, and nobody had found 'em.[3]

Kathy: Cause they were in the middle of the river . . .

Greg: If you go to the bow of the *Tremble*, you're actually a hundred yards offshore, which is quite a distance when you feel you can't come up (you have to come back to shore to come in). Well that was my starting point for the dive. I would race as fast as I could out to the bow of the *Tremble*. So I'm a hundred yards out. And then from there the dive started. I'm thinking it's probably another hundred yards to the place I found.

I was coming up the slope of the other side of the river. So I was in forty-five foot of water, and this downbound thousand-footer snuck up on me. That was one time I thought I was really gonna be in trouble."

I was on bottom, but I was at forty-five feet.

Kathy: He was shallow. He's in forty foot. If it's drawing thirty, now you got ten foot of clearance between you and a huge freighter blade, spinning, and they do draw. That's not much margin. It went dead over him.

Greg: It's a suffocating feeling. You think, something that big: I've never seen anything as huge as that underwater. I had great visibility, and I'd never been that close to a freighter, and my thought was, "I could just go up and touch that." I mean it was that close, it was like, if you were in a swimming pool and going up twelve feet to the surface. I thought, "uh, this is it."

So I looked around.

Kathy: You could see the rivets. You've got a thousand foot to figure out what to do before the blade gets to you. You can't go deeper, there is no deeper.

He was on gravel bottom. No hand-holds.

Greg: I did find one bent pipe, and I went over to the bent pipe, and the holes were all zebra encrusted, and I just jammed my fingers into the pipe through the zebra mussels, and I put my head down on the bottom because, I just wasn't sure, I'd never been that close.

Kathy: If they're that heavy and that shallow they can draw off the bottom. He's thinking if it's spinning rocks, if there's all this gravel flying around, if it breaks his mask, now you're in serious trouble. It's something as simple as that. He might not actually get sucked into the blade and chopped up, but if it smashes his mask and floods his mask now he's in the middle of the river under a freighter with no sight, cause then you're blind.

Greg: It's the strangest feeling. My first thought was, "I'm gonna be crushed." The boat had to be wider than this building. It's all just a mass above you. And now it's dark. The only light is over there, and way over there, in either direction, and you're right under there. It's the weirdest, weirdest feeling the first time you get caught like that . . .

Surface

Kathy: *So if you pop up . . .*

The most deadly to divers are your sailboats and windsurfers. That's why we never wanna come up. Because if you're on the surface, even if you're surfacing, and a thirty foot sailboat with a six to eight foot keel comes by, you're dead. You're not gonna hear it, and it's gonna take you out so fast you probably won't even know what happened.

Greg: I almost got runover by a sailboat in the British Virgin Islands when I was snorkeling. Yea, that was, that was more terrifying than a powerboat cause, he almost hit me and I never even heard him. Didn't hear him coming.

Kathy: Even a windsurfer—we have tons of windsurfers up here—I mean they're flyin.

Greg: Oh yea, they're movin.

Kathy: I don't even know, they could be going twenty miles an hour.

Greg: Oh yea, at least that.

Kathy: I don't even know how fast those guys are going. They are flyin.

So if you pop up, the first thing that pops is your head. So you pop your head out of the water, you get hit, it's a closed-head injury. Literally you're never gonna know what happened to you. You're just gonna be dead.

See we don't look at that as a bad thing per se. Beats gettin hit by a drunk driver. But it's going to be as immediate.

Greg: Yeah, hopefully.

Kathy: Yeah, hopefully.

Soundscapes

Greg: *It's noisy as heck down there . . .*

Kathy: You're always hearing that freighter thump in the background real loud. And then the boats: You can hear the difference between power boats and jet skis and motor boats. They all have different pitches. The smaller the engine, the more whiny it is. *Bzzzz.* The cruisers, let's say thirty footers to sixty footers, they're quieter. They're not doing the thumping, and they're not doing the squealing.

But the motor boats (the fishermen) and jet skis are really loud. They're like *bzzz* [lower pitch], and some of 'em are like *bzzzz* [extra high pitch], and them some are like *bzzzz* [medium pitch]. That's just constant. So there's constant buzzing.

Lynne: Do fish make sounds?

Kathy: They ruffle the gravel. When the female spawns she grriinds her belly in. She wants to get those eggs as deep into the gravel as she possibly can. And that's what we really hear, the grinding. It's very distinct.

You can hear it just as soon as you stick your head underwater. Like up at the shore: even if the fish are out in sixty, seventy, eighty foot of water, spawning way out there, a football field away, you can hear it.

I hear a lot of crackling and popping, like from shrimp. It might be things shifting. Everything's kind of alive.

Greg: I'm more aware of that in the ocean. Jacques always said it's a silent world. I mean it's noisy as heck down there. There's shrimp cracklin all the time. I don't hear that like you might hear it in the river because you've got better hearing.

Kathy: Yeah, in the river, I hear it. Greg doesn't hear the low tones anymore. I think it's from all the years of equalizing. It puts a lot of pressure on the ear drums.

Greg: Most divers seem to have not very good hearing.

Merganser

Kathy: *Where in the hell are they finding these fish?* . . .

Greg: So it's winter, you go down, you don't see a fish, you just don't. You might see a sculpin, if you're by a discharge that's warmer. You might see a sculpin.

So, we're at a friend's house. It's a little thing right on the river. And we're at a friend's house, and it's winter. And diving ducks all over the place. A lot of them eat seaweed or moss or whatever they eat. And some eat fish. So, see a merganser go down and, I was just in that spot the day before. There are no fish down there. That merganser dives down, and he's got a fish within a minute.

Kathy: A minute! I mean a minute! I mean he pops up, and you can see it, he's throwing it around in his mouth like a penguin or whatever. We're like, "what the hell?" For a whole hour they did it. One after another—boom, boom, boom, boom, fish, fish, fish, fish, fish, fish, fish.

What the hell, where in the hell are they finding these fish?

Greg: So I go back in, and I'm pretty sure they're eating gobies, and you can't find a goby in the winter to save your soul, and so, you gotta figure they're under rocks. I'm turning over rocks to see if there's gobies hibernating underneath. No. But if you get a big rock, that's really in there and hard to move, and some of 'em it's all I can do to move these rocks even underwater, there'll be four or five gobies. All's their eyeballs are sticking out and their gills and everything are under the rock. You can grab 'em and they'll wake up a little bit. You can just grab 'em.

So those ducks are moving not just little rocks . . . they're movin rocks that are *big*!

Kathy: Boulders.

Are they doing it as a team? Did they work a hole and now they can just go in and get it? We don't know, but we wanna film that so bad. Ducks are the smartest, my god.

Minnow

Greg: *The hornyhead chub . . .*

He'd fix the nest a little bit and he'd spawn some more. And then when he was all done, he went and, picked out rocks and, as big as his head he'd pick it up with his mouth and he'd come over and he'd look and he'd plunk it there.

Kathy: He buried the whole entire thing. It was really cool. He buried the whole site.

Solo

Kathy: *Didn't feel like saving anyone today . . .*

Every time Greg's at the river people ask, "Why are you diving alone?" That's the only thing they know about diving is that you're never supposed to dive alone. That's a rule of diving. But that's sport diving. We dive commercially. Commercial divers never dive with someone. They're always alone. It's completely opposite. So once you transition to professional and commercial diving, you leave your buddy behind. The whole idea of a buddy goes away.

But people don't understand. He's tried all different kinds of comebacks and spunny things and explaining to 'em. And people have argued with him. They don't know anything about diving. They're not certified divers.

So now you generally say you didn't feel like saving anyone today.

Walkabout

Kathy: *He has a visual map of the entire river bottom . . .*

Greg: The dive out to the *Martin*, which is about to the bridge and has all the iron ore on it, that's probably the most advanced dive in the river.

You have to get in at a certain spot, which is the drain from the roads, the drain pipe.

And then you gotta go through the shallows, and go past all the redhorse suckers, and then you get to the first drop off, and that's the first picnic table you run into, and then you grab the rope that's hooked down there for some reason, and you've gotta get above the big rock, and then the idea is to kick for Canada now.

And hopefully you'll hit your landmarks, and my landmark is the big twelve-inch pipe that's there, then you're just over rocks, but if you keep going, almost every time you hit the chain-link fence that's in a pile. You always stop there and get your breath.

Then you go again and you hit the log that's got another rope that's tied on for some reason, and you can grab that, and then you can hold there again, and then you can pivot the rope, kinda turn your shoulder into the current, and just have the current pivot you as far over to the outside as possible, and then you go again.

Let go and start kickin again, and, if you miss the second picnic table, from the beach, if you miss that, or at least if you're on the downstream side, then you're goin down in the trough, cause you always end up in the trough, and you go down the trough, and that's where you're gonna be pushed down to eighty-some feet, and you just ride that trough.

If you wanna stay (it depends on what you wanna film, I usually want to film something when I go out there), you duck in behind that picnic table and go there, and then follow the wreck downstream. But a lot of times you go down in the trough, and there's this other real huge pipe that's there, and you really have no control. You're just forced straight down there so your ears better equalize, and you just get down there, and once you hit the bottom of the trough, well then there's a big slope up to the wreck, and it's probably twenty, twenty-five feet of wreck and debris, and you get down there, and then you can decide where you wanna go.

You can go back upstream. Or if you just wanna head to where all the walleye hang out, then you go downstream a little and you come up over all this debris that's gathered over the years. Then you hit the wreck, the actual hull of the ship, then you go back around, and then there's this sand spot that's usually nothing but walleye. Just hundreds of walleye hang out there.

It depends on the time of year, but, as you're being forced down in that trough, a lot of times you'll see the sturgeon there, and they'll get out of your way cause they can tell you have no control of the situation.

Homeward

Greg: *You hit what we call the elevator . . .*

Depending on your air supply and what your dive plan was, but if you wanna come back to the parking lot you parked in, you have to go back up the wreck, underneath the hull, and you follow that. It's very protected, very calm right there. And you go by all the catfish in there. You go up, and then you end up at the picnic table, and the trough. And if you go in right there, and cut across the trough real quick, and go up, you hit what we call the elevator. That's a current that will take you from that area and put you right at shore.

Kathy: Straight. You don't have to kick, you don't have to swim, nothing. It's a weird current. It's almost always there. It's a rip current or something.

Greg: Perpendicular to the river current. The current's going north and south, and this is going east and west.

But if you leave the wreck from the downstream side, you have to have a lotta air, cause the current is so fast that as you leave the wreck, you gotta go down into the trough. Well, you can't always go through the trough down there because there's one current there that'll take you to Canada.

Kathy: And it's partly because of the wreck. The mouth of the river comes in so you've got all that velocity and you've got a wreck that sits right there. A lot of stuff gets hung up on that wreck. It has scoured out this trough that we're referring to. Main flow is going crazy past there, and then you've got these two offshoots . . .

Greg: The one to Canada, it's kinda scary because, if you know you wanna go that way or that's what your dive plan is, then you have the air and everything to do that. And that would be an offshoot of how you would do a rock dive to get the Indian net sinkers. You would purposely go to the end of the wreck and let the current take you. And then you're right out in the center of the river and you're goin under the Blue Water Bridge, and there used to be a lotta good rocks there.

But if you wanna come in from the end of the wreck, it's a long journey in, and you gotta have a lot of air, and you kick and you kick and you kick and you kick. And eventually, you will get to the zebra mussel line, which is a real indicator. On the river, there's clean gravel—we call it clean gravel—and that's down the center. Nobody's been able to really colonize the center so much, of the clean gravel, which is good for us.

Then there's a zebra mussel line that is almost like night and day.

Kathy: Like the shoulder of the road.

Greg: And that's a real good thing when you're waaay, way out.

Kathy: We estimated it's as narrow as maybe twenty or ten foot in some areas, and it can be a hundred foot wide in others. It's like a ribbon. And then you get into the near-shore zone.

You love to see the zebra mussel line cause you've made it then.

Greg: You know you're gonna live at least. "Ah OK, I'm at least gonna make it." Makes you really wanna go divin with us, doesn't it?

Appreciation

Greg: *Look where you are . . .*

When I'm going to get out of the water I take my fins off and I put 'em on my wrist. I'm sittin there and I always look around and think, "look where you are," cause it's usually really nice vis., and so I always say "look where you are."

And then all's I do is, very slowly, crunch, crunch down, and then spring as hard as I can up to the ladder. I fly up twenty feet and grab the bottom of the ladder, and then slowly come up.

I usually do the same thing when my head comes above water: I look around from water level to see and think, "look where you are, look what you just did."

I always try and appreciate that feeling when I'm under there.

(Seeing + Knowing) × Time = Hope?

At the end of his book *Landmarks*, British nature writer and explorer Robert Macfarlane has a short chapter called "Childish." *Landmarks* should be required reading for anyone fascinated with the languages of place. "Childish" should be required reading for anyone who sees their self as a wildly curious person.[1] In "Childish," Macfarlane introduces us to biological anthropologist turned early childhood educator Deb Wilenski. Wilenski once followed a class of children into parklands near their English hometown. Later she shared their wisdom and mind-blowing map making in *Fantastical Guides for the Wildly Curious: Ways into Hinchingbrooke Country Park*.[2] As Macfarlane synopsizes, Wilenski wanted "to record without distortion how the children 'met' the landscape, and how they used their bodies, senses and voices to explore it."[3] Left alone—no teachable moments from adults allowed—the children found "a limitless universe." They became inventors of place. But use an adult word like "place" with care, says Macfarlane. "What we bloodlessly call 'place' is to young children a wild compound of dream, spell and substance: place is somewhere they are always *in*, never *on*":

> The hollows of its trees were routes to other planets, its subterrane flowed with streams of silver, and its woods were threaded through with filaments of magical force. Within it the children could shape-shift into bird, leaf, fish or water.[4]

Wilenski's mapping project summons the spirit of Theodore Seuss Geisel—Dr. Seuss—an earlier depicter of childishness in nature. Before *One Fish, Two Fish, Red Fish, Blue Fish* or our environmental anthem, *The Lorax*, Dr. Seuss dedicated a book called *McElligot's Pool* to his father, "the World's Greatest Authority on Blackfish, Fiddler Crabs and Deegel Trout." According to Jack St. Rebor's delightful *Seussblog*, "the 'deegel trout' is a private joke referencing some of the Geisels' more unsuccessful fishing trips when his father would purchase trout from the Deegel hatchery and pass them off as their catch."[5]

The book's protagonist was a young visionary, Marco, who was fishing in a lowly junk-infested pond when a skeptic approached.

> *"Young man," laughed the farmer.*
> *"You're sort of a fool.*
> *You'll never catch fish*
> *In McElligot's pool!"*

Under dour critique the boy was still generous. "Hmmm . . ." answered Marco, "It *may* be you're right." But Marco didn't see the pond in shades of hum-drum. Wonder awaited.

> *This MIGHT be a pool, like I've read of in books,*
> *Connected to one of those underground brooks!*

For Wilenski's children in Hinchingbrooke Country Park, perhaps the converse of their inventing also happened. Perhaps the same beings and landscape features with which the children animated their found places—trees, birds, fish, ponds, puddles, streams, paths—had been inventing and animating them as human beings. In this layered land of childish place-making that amazed first Wilenski and then Macfarlane, adult lessons could be found:

> *Learning had primacy over factoidal information.*
>
> *Intellect was indistinguishable from imagination.*
>
> *Humans were symbiotic with landscape.*
>
> *And work was—most energetically—play.*

———

"My number one reason for jumping in the water and going diving," says Kathy Johnson, "is to see and play with the fish."[6]

For Kathy, the St. Clair River is a world-class aquatic playground. As a corridor between the upper and lower lakes, most Great Lakes fishes swim in the St. Clair. On one dive, Kathy and Greg might see rainbow darters and sculpin, muskie and sturgeon, six-inch logperch and thirty-inch bass, walleye, steelies, even an eastern spiny softshell turtle.

A steelhead trout once gave Kathy the most enchanting moment of all her years exploring the St. Clair. "It was the first time I ever tried to feed a steelhead," she recalls. She brought brine shrimp and bloodworms to entice a fish. One took it right from her hand. "This guy was really aggressive and he just fed like crazy and he was so beautiful and he was huge. I'll never forget."

Steelhead is also a sport fish in the St. Clair. "You do realize that you're feeding the appetizer to the main course, right?" asked a not-so-enchanted angler. "Yeah," Kathy replied with a smile, "in your world." But below, feeding the steelie, "that's the ultimate trust," she muses. "You've broken that barrier of wild animal and person."

"That's my favorite thing to do, and I can spend a whole dive trying to get one fish."

Kathy has urged dive shops to add shallow water locations to their tours, along with shipwrecks. Guides told her customers wouldn't be interested, but Kathy is sure that direct encounters with a *living* river would be irresistible. Put people on these sites to photograph fish. Put Marco in a wetsuit, and what would he discover?

—⌒—

Kathy is wildly curious. So is Greg. They could write their own *Fantastical Guide for the Wildly Curious*. Somehow they sustained their curiosity through decades of troubles besetting the Great Lakes and their own persecuted St. Clair River. Somehow they stayed childish. Could it be that *in* the St. Clair they, too, found "a limitless universe?" That, in Macfarlane's words, "how they used their bodies, senses and voices to explore it" gave them rare sight and still rarer knowing? That the river itself carried them along counter-currents of unconventionality, away from any main*stream*?

Consider zebra mussels, quaggas, and round gobies: Greg and Kathy look at them differently today. They filmed the river in the years when round gobies got established. Kathy plays some of that footage. Minute after long minute, the only fish onscreen are gobies. *One fish, one fish, one fish, one fish.* "We really did think we were going to lose our darters and our logperch and our sculpin. All of our bottom fishery."

Twenty years later, they went back to that spot to shoot, and gobies weren't so prominent. The two continued to visit, and now they see bass, schools of logperch, even sculpin. Early on sculpin "took the biggest hit," Kathy explains. Now "sculpin are good," affirms Greg, "I see a lot of sculpin."

Their re-photography is an exposé of hope. First there's the dreary "before" imagery, with its all-goby monochromatic repetition. And then the "after," not fully Kodachromatic like you might see on a tropical reef or in *Finding Nemo*, but still a living collage that swims, schools, scatters in and out of view.

Kathy thinks back on that early footage, right after goby populations exploded. For a while, the new fish had no predators. "Nobody was eating gobies, because nobody knew to eat gobies."

"It took the bass," she says. "They needed to figure out that gobies were a food source. And the second they figured that out, they just went crazy."

"So now, in the St. Clair, where we are right now, bass masters are coming from all over the country, because we have record breaking bass. I mean huge bass— we're seeing thirty, thirty-six inch."

"And when we dive they follow us around like packs now. We call them huntin packs of bass. We disturb rocks and stuff, so they follow us. They've learned, and they know the second a diver hits the water. I take my hand and I hover it over a rock, and they're like, 'Wait!' and they zoom to get downstream cause they want to be in position downstream of the rock when it lifts. When I lift it they wanna be able to go in at the fish, and if the fish goes upstream they can take it, if it comes downstream they're in position, and if it goes darting to the side they got it."

"I mean they're literally like 'whoo-whoo.'"

"And it's funny," Greg says, moving from bass to walleye, "because the walleye, which I would think wouldn't eat a goby if its life depended on it, now every fisherman cuts open the stomach to see what it's eating, when they filet the fish, and usually you'll find three gobies inside their stomach."

"A walleye isn't really built for that. They're more for minnows, you know? They're streamlined and a submarine and they can shoot after 'em. Yeah, bluewater hunters. But they're eatin, they're eatin gobies."

Kathy says the tribes are finding the same thing with whitefish. Is there too much hype about invasive species, she wonders, "the henny penny thing?" She notes that at their re-photography site, "there's no gobies visible on the bottom whatsoever."

"Now they have to hide," Greg says. "They have to hide," Kathy echoes. "They're back to their natural behavior. So their population is totally controlled."

The two send other good news from below. Near Port Huron's shore, Gregory A. D. recently discovered refugia of native unionids, including a bed with five species. Biologists had been unaware of unionid remnants in the upper part of the river. They even spotted a black sandshell in the Gregory A. D. footage. *Ligumia recta* (Lamarck 1819), the black sandshell is endangered in Michigan and rare in its historical Huron–Erie range. The bed lay within Section 6.0 of Project 3 of the Upper St. Clair River Restoration Project. With its discovery, the St. Clair River BPAC called for "a full natural feature inventory, both on land and under the water, before . . . restoration projects begin."[7]

Kathy also cites the research of unionid scientist Dave Zanatta. Zanatta had spoken about remnant freshwater mussels at a symposium hosted by BPAC called "The St. Clair River: Bridging the Environment and Economy." The symposium press release said it would "provide the public, industry representatives and volunteer organizations with an excellent opportunity to hear a diversity of experts from both sides of the border who are working to improve and protect water quality in the St. Clair River."[8] Post-symposium, Port Huron reporter Jim Bloch followed up on Zanatta's presentation with this headline, "Written Off as Doomed, Native Mussels Survive Zebra Mussel Invasion."[9]

The US Fish and Wildlife Service calls unionids "silent sentinels."[10] Of all the wildlife in North America, unionids are the most endangered. Thus they are the silent sentinels of our extinction crisis. Zanatta didn't downplay how unionid remnants sit on a knife's edge of survival. But one of his research articles confirmed that, "the St. Clair River Delta has persisted as a refuge habitat for native unionid mussels."[11] In another article, he reported that "impacts of zebra mussels on unionids are most detrimental during the first stages of invasion but that impacts diminish over time."[12]

At this point, Kathy is about to engage in some serious Great Lakes heresy, which she prefaces with, "We're trying to get the message out to places like Traverse City."

〜〜〜

Two-hundred-fifty miles northwest of Port Huron, Traverse City, Michigan, sits on the quartz-sand coast of Lake Michigan, where the Boardman River empties into Grand Traverse Bay. Nearby are the dunelands of Sleeping Bear Dunes National Lakeshore, whose perched dunes soar then fold with the wind, which works them into crystalline waves and wrinkles. The endangered pitcher thistle relieves glare from the sand with silvery-green tufts and tendrils. In scoured bowls between dunes, a few inches of water accumulate in ephemeral pools. These interdunal wetlands create oases for plants, insects, birds, and amphibians. They teem with unnoticed life as people rush past to reach the opal sea ahead. At the shoreline of the big lake, everyone slows to a stroll, their own ephemeral prints pooling as waves hit hard wet sand. Sometimes footprints sidestep long mottled ribbons of empty zebra or quagga mussel shells.

Like Kathy and Greg in the St. Clair, a local videographer has filmed mussel colonies in Lake Michigan. Hans Van Sumeren directs the Great Lakes Water Studies Institute at Northwestern Michigan College. He's not a diver, but he is a world-renowned innovator of marine technology and expert with deep-water remotely operated vehicles (ROVs). Hans sends me underwater footage from Grand Traverse Bay, September of 2013, two hundred to three hundred feet deep. Hit play and the journey begins.

The drone hurtles downward through a snowstorm of particles in the dim water column. The scene jerks right then left, off kilter, mimicking that 1990s hand-held effect of a low-budget horror flick. Maybe this is an underwater *Blair Witch Project*. The only noise is a *whooshshshsh* that whistles and hisses in two sound layers, arctic wind and snow-blower. (Unfortunately, muting the sound while viewing is disturbing in its own way.) Pointing the GoPro camera clamped to his ROV, pilot Hans selects substrate to survey. He brings the scene into focus. It looks like we're about to land on a rocky planet whose singular lifeform is mussels—cities of mussels, forests of mussels, fields of mussels, a monocultural underworld. *One fish, one fish, one fish, one fish.*

Traverse City is earlier into their "introduction," as Kathy calls the initial invasion of zebra and quagga mussels and the round goby. She commiserates with regional fears for native fishes in Lake Michigan. But she hopes that agencies won't try to combat mussels or gobies with technology—she'd heard sonar or sound barriers. "Hang tough" is Kathy's heresy. *Of course* technology is tempting. (You can't influence quaggas with social media or mass protests, and people want the relief of doing something.) But like the introduced species themselves, introduced technologies might pose threats. "There's so much we don't know about impacts on other fish, including lake sturgeon."

"Just let it be, and let it balance out," soothes Kathy, aquatic yogi, exhaling sanguinity.

Look to the St. Clair River delta when it was ten years in, "when the bottom was just nothing but zebra mussels and gobies, solid." Kathy and Greg saw food-web adaptation in real time, and later, thirty years in, even a cautious scientist like Dave Zanatta was willing to tell a reporter that "there was reason for hope."[13]

From Kathy to us: "Our culture is so immediate, right? Everyone wants a solution now. It's like, give the fish a chance, ya know? I mean they'll eat gobies. Turns out they love gobies. They just had to figure out they liked 'em. And it took 'em a couple generations. Fish live five to seven years, so you need to give 'em ten to fourteen years."

"Give it another ten years."

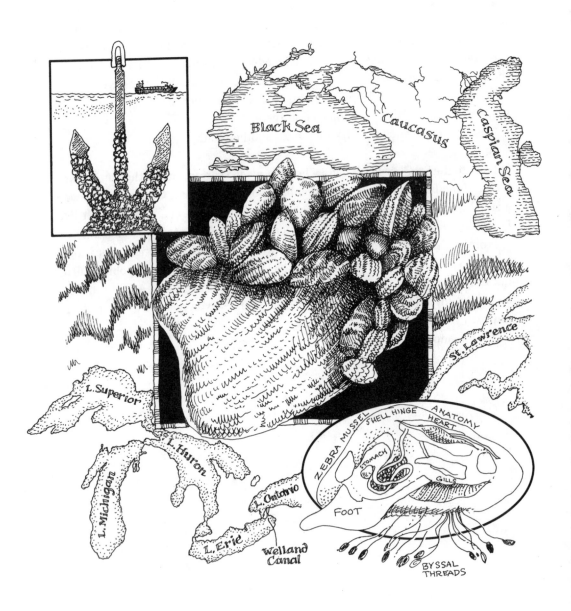

Black Sea

Caucasus

Caspian Sea

L. Superior

L. Michigan

L. Huron

L. Ontario

L. Erie

Welland Canal

St. Lawrence

ZEBRA MUSSEL SHELL HINGE ANATOMY

HEART

STOMACH

GILLS

FOOT

BYSSAL THREADS

Part III. The Paradox of Abundance

OR, PROBLEMS OF SCALE

Negotiating Abundance and Scarcity

with Daniel Macfarlane

In 1982, a collective chill spread through the offices of two Canadian premiers and eight US governors whose provinces and states encompassed the vast Great Lakes–St. Lawrence basin. In *Sporhase v. Nebraska*, the United States Supreme Court had just declared water an article of commerce subject to interstate trade under the commerce clause of the US Constitution.[1] Henceforth states could not ban water diversions outside their borders, the question addressed in *Sporhase*. Imagine the implications from the perspective of policymakers and politicians in Great Lakes states. At six quadrillion gallons, 84 percent of North America's surface freshwater, the lakes were a kind of aquatic El Dorado, hypothetically open to those with the political or economic might to extract their water.

Such fears were not hyperbole in the Great Lakes. The court decision came on the heels of a US Army Corps of Engineers study on whether imported water, possibly from the Great Lakes, could restore a rapidly declining Ogallala Aquifer in the Great Plains. The corps provoked more paranoia by using uncharacteristically socialist language about fairness, or redistribution from "water rich" to "water poor" regions.[2] Proposing a national water policy became a kind of shorthand for water redistribution.[3]

The corps study followed the resurrection of an infamous proposal by Canadian engineer Tom Kierans. Kierans named his idea the Great Recycling and Northern Development Canal, or GRAND. GRAND would pump water to Lake Huron from James Bay, which lay far to the north on the southeast corner of Hudson Bay. According to the GRAND concept, Lake Superior would no longer be necessary to feed Huron. Therefore a channel could run Superior's "superfluous" water to the arid—i.e., water poor—American West. These and other epic ideas raised the hackles of whichever Great Lakes premiers or governors were in office.[4]

The *Sporhase* case galvanized an intense twenty-five-year saga of interstate, interprovincial, and binational negotiations with one goal: to find a constitutionally sound and mutually agreeable way to limit future diversions. After many twists, turns, setbacks, and advances, in December of 2008, a binding Great Lakes–St. Lawrence River Basin Water Resources Compact took effect (hereafter the Great Lakes Compact). The compact and its companion Great Lakes–St. Lawrence River Basin Sustainable Water Resources Agreement with Ontario and Quebec put limits on water use and diversions from the basin. This was an environmental landmark of the twenty-first century.

With the compact in hand, stakeholders from local to federal levels seemed to escape what we might call "the paradox of abundance."[5] Many environmental histories share an abundance narrative—i.e., that the intense concentration of a valuable resource practically assured the decimation of that resource. The historical reasons vary, but for the nineteenth and twentieth centuries, reasons often involved time lags between market-driven extraction, increased scales of production, catch-up policy responses, and true care for the natural world. Such boom-and-bust histories along today's Canada–US border include near-extinction of the beaver in New France and bison on the nineteenth-century Great Plains, and the actual extinction of the passenger pigeon; liquidation of old-growth white pine forests in the Laurentian Great Lakes; fishery crashes from the Grand Banks to the Great Lakes to the Pacific Northwest; mineral mines, including gold strikes along the Alaska–British Columbia border; and Canadian oil and especially the infamous Alberta tar sands crude, much of it sent south across the border.[6] Abundance stories are Sisyphean: our economic and cultural inability to prevent the next example, to push the proverbial rock over the crest, to *sustain* both the people and the nature of our homes.

In North American environmental history, abundance is a powerful narrative indeed.[7] But in North American water history, scarcity is the dominant narrative.[8] The western half of the continent—the American southwest especially—has had an understandable but nonetheless disproportionate influence on national narratives of water and debates over policy. Think of "the border" itself. For many Americans, and certainly the media, the first border that comes to mind is the US–Mexico border and its borderline through the desert, the Rio Grande.

Historically, think of American John Wesley Powell's explorations of the Colorado River in the nineteenth century and his unheeded recommendation that climate-appropriate property boundaries should restrain settlement in arid regions.[9] A decade before Powell, geographer John Palliser made nearly the same argument about semi-arid dry prairies in southern Saskatchewan and Alberta. Much later, irrigation transformed "Palliser's Triangle" (the area's common name) into Canada's breadbasket of wheat production.[10]

While scholars, policymakers, and environmentalists still look to Powell's journals for insight, a legal system of water rights at odds with his approach prevailed in the arid American West. The system's bulwarks were the Colorado Doctrine, governing individual user rights, and the Colorado River Compact of 1922, an agreement among the river basin's seven states to allocate water rights to the river and its tributaries.[11]

Better known as the prior appropriation doctrine, or "first in time, first in right," the Colorado Doctrine separated water rights from riparian land ownership.[12] In simple terms, prior appropriation means the first user has the superior claim to a water source. This claim holds even if later users own land adjacent to the water and the first user owns no adjacent (i.e., riparian) land. The key is that the first user's purpose be "beneficial," which historically meant for agriculture or industry. For instance, if the first user was a mine operator who diverted water from a stream to run the mine, a second user could not interfere with that first use. After the first user, the second user had the next highest claim, and then the third user, until, theoretically, there was no water left to use. Prior appropriation made water a quantifiable and transferable commodity; therefore, a user could divert water to another location and sell his user rights and legal place in line to someone else.

Today, both prior appropriation and the Colorado Compact are broken. In an epoch of global warming and megadroughts, there is not enough water to share but still enough to fight over. "Colorado to California: Hands off Our Water," shrills a Fox News headline.[13] "Rain Barrel Bill Dies on Calendar," runs a much blander headline in the *Colorado Statesman*, though this 2015 story was just as dramatic. "A bill that would have allowed Coloradans to collect rainwater died in the Senate late Tuesday night," begins the rain barrel story—and midway through is the crux:

> Opponents, including farmers and ranchers, believe that rainwater is covered under the state's prior appropriations law, since it runs off into groundwater and surface water, such as rivers. . . . There's a reason why rain barrels have been illegal in Colorado for the past 160 years, according to Chris Kraft of Fort Morgan, who operates one of the largest dairies in the state. "We're short of water. People keep moving here. This is a worse idea today than it was a long time ago." Kraft explained that farmers have to get a water court decree to get water, and some of those decrees date back to Colorado's earliest days as a state. Kraft said his decree dates back to the 1890s, and he has to pay a lot of money for that decree and the ditch that supplies his farm with irrigation water. "This would allow people to steal water from my appropriation," he told the Ag Committee.[14]

In North American cartography, the one hundredth meridian is a legendary signpost for climate. The one hundredth meridian marks the transition from arid to humid lands. To someone who lives east of the one hundredth, "rain barrels" don't sound like fighting words. That collecting rain from one's roof has been illegal anywhere might be a stunning idea for, say, a Michigander or an Ontarian. In more general terms, however, popular culture has made conflict over scarce water a Pan-American narrative. In the famous 1953 western film *Shane*, ranchers and homesteaders warred over land with access to water. As they fought, the story goes, a moral code and rule of law emerged to civilize the American West and point the country toward greatness. No matter that the 1950s parable about the 1880s frontier was belied, even then, by the 1930s Dust Bowl.

With its prominence in American politics, literature, and film lore, scarcity dominates how many of us see water. Iconic images of Dust Bowl suffering and a new iconography of water scarcity are bookends to more than a century of dry-land visuals.[15] From *National Geographic* to local newspapers, twenty-first-century photos of cracked landscapes make water the focal point by its absence. Often a dark line leads the eye through the parched scene—the S-curve of a bone-dry streambed.

All of this raises a question: If much of the history of the American and Canadian West is variations in the key of water, why is there no equivalent filmography or literature or iconography for the Great Lakes region? Surely its history includes an awe-inspiring water narrative? Surely its immensity as the largest freshwater system in the world could rival the immensity of water scarcity out West? But we wager that the average Coloradan gets little exposure to the Great Lakes through education, political discourse, or the cultural imagination. Author Jerry Dennis once marveled that the Great Lakes are so unknown beyond their shores that a funny online hoax about whale-watching in Lake Michigan made its way into a children's K–6 science magazine. A Michigan teacher had to alert the publisher's editorial staff in Utah that, no, whales and dolphins do not set forth each spring from Hudson Bay to breeding grounds in Lake Michigan.[16]

Dennis hypothesized that people do not "see" the Great Lakes because the lakes are too enormous and diverse to comprehend. Yet the West is enormous and diverse, too, on both sides of the border. So we'll add two other hypotheses. First, perhaps their low visibility in water discourse is because the Great Lakes make up the actual border between the United States and Canada. Their significance cannot wholly fit nationalist narratives of development and identity, and their governance is easily banished to the far-away realm of diplomatic niceties or border checkpoints, rather than the knock-down, drag-out arena of the rain barrel. By contrast, the upper Colorado River is a wholly US example. As such, even Easterners might see a battle between rain barrel friend and foe in more familiar terms, as the latest local resource controversy to intersect with state or national politics.

For our second hypothesis, the Great Lakes might fade into another kind of distance—emotional and empathetic distance, or the degree to which people can imagine themselves in a distressing scene. A few ugly invasive species or an economic legacy of industrial pollution might not trigger the same empathetic intensity or emotional visualization from outside the region as the apocalyptic specter of two countries' breadbaskets disintegrating into dust while scientists forecast the inexorable drain of ancient aquifers like the Ogallala. Perhaps water scarcity west of the one hundredth meridian mapped a sharper, more dangerous geography in the public imagination than do water regimes east of the one hundredth meridian, even someplace as physically distinct as the Great Lakes.

Nonetheless, we might have reached a turning point. A geography of water abundance—one in which Utah textbook writers could picture make-believe Lake Michigan whales—now includes its own all-too-real, fully imaginable site of empathetic horror: the water crisis in Flint, Michigan. A conspiracy of negligence that lead-poisoned a city became, if possible, more terrible because Flint residents once had, were recklessly deprived of, and yet remained painfully close to abundant safe water.[17] Flint generated a new emotional Great Lakes geography that momentarily transcended politics, occupation, class, and color. How easy to imagine yourself in a Flint home whose water tap holds invisible terrors and irreversible harm for your family. On this mental map, the home is only inches from Lake Huron, the fourth largest freshwater lake on Earth.[18]

We argue that alongside scarcity, abundance has been a different but powerful driver of water law, policy, economics, and culture.[19] To give one abundance example from the same frontier period when western states and provinces were experimenting with laws on prior appropriation: far to the (humid) east, the state of Michigan, surrounded by four of the five Great Lakes, established a matrix of laws and property rights to drain water from as much land as possible.[20] "Don't go to Michigan, that land of ills, the word means ague, fever, and chills," warned a nineteenth-century chant about the state's reputation as a swampy, disease-ridden hellscape for settler farmers.[21] The culmination of Michigan's exertions to deal with surfeit or "too much" water was the Michigan Office of Drain Commissioner, a county-level elected position that some political scientists uphold as a candidate for the most powerful local elected office in the United States or Canada—or, "the state's most powerful man," according to a boastful Shiawassee County drain commissioner in 1979.[22]

The 2008 Great Lakes interstate compact and its companion binational agreement raise another important question. The compact marked a partial reprieve from boom-and-bust water exploitation. So far, at least, the basin is not on track to slake an insatiable dryland thirst, or become a liquid mine for twenty-first-century robber barons, or, scariest of all abundance nightmares, shrink into a poisonous salt barrens from economic hubris, like Russia's Aral Sea.[23] Why did the Great Lakes escape this paradox of abundance? With an international maritime corridor, with a withering industrial base (steel, chemical, paper, automotive), and with aquatic ecosystems compromised by toxic pollution, invasive species, shoreline development, and climate change, it seems remarkable that eight American states, two Canadian provinces, and two nations could come to an agreement on a legal, economic, and environmental matter as contentious as controlling water.[24] Yet they did.

Note that the question is not how the region escaped the paradox of abundance. The how is part of the history of negotiations. But *why?*

Was it because the basin's state and provincial governments were somehow more evolved than their brethren to the north or south? Were they more virtuous, more altruistic, maybe smarter than their counterparts along the Colorado River? Hardly. (However, far be it for us to assume Mark Twain's mantle of political "moralist in disguise.")

Was it because water itself was such an exceptional resource, fundamentally different than fish, trees, or ore? No again—at least not legally. To the contrary, many warned that the compact enshrined water as a commodity and carried unfortunate echoes of the prior appropriation model. Critics like Dave Dempsey argued that policymakers compromised away a strong constitutional case that Great Lakes water should be subject to a public trust doctrine instead.[25] The public trust doctrine traced its roots from ancient Roman civil law to English common law and ultimately to a robust body of law in the United States—both in the states and nationally with affirmative Supreme Court decisions.

Was it a higher moral imperative that outweighed other considerations? That water is so fundamental to human and nonhuman life in the region that their welfare demanded it be protected from outside claims? One might hope so, but again, no. In fact, the moral argument often went against protection. In a world where billions of people are without potable water, how can you win an argument against urgent care for your brothers and sisters? The short answer is you cannot win that particular argument.

So, why the good outcome?

We think one explanation lies in the US–Canada border itself.[26]

———

Borders embody dualisms: they divide yet potentially unify, they are barrier yet possible gateway, they are solid (on paper) yet porous, they can intensify competition or inspire cooperation, they can stir resentment or nurture understanding. Borders are complicated.

International borders are even more complex. They are actual places, just as regions and provinces and states are places. International borders can loom large and brooding in a nation's political consciousness, as the Canada–US border does for Canadians and the Mexico–US border does for Americans. Or, they can recede to the edge of a Rand McNally atlas, as the same US–Canada border does for many Americans.

Border waters complicate things still more. For Canada and the United States, shared waters were more than a river delineating two countries, like the St. Lawrence or Detroit or St. Clair Rivers. They were more than a major river crossing two countries, like the Columbia. The forty-ninth parallel between our two countries includes 2,200 miles (3,540 kilometers) of boundary waters, from the Bay of Fundy on the Atlantic to the Salish Sea on the Pacific. To the north, the border continues between Alaska and British Columbia.[27]

List these border rivers and lakes, and you'll find signposts to great swaths of North American history and geography: in the northern reaches, the Yukon, Chilkat, Stikine, Taku, Firth, Whiting, and Alsek Rivers; along the southern Canada–US border, Columbia, Skagit, Kootenay, Pend D'Oreille, Flathead, St. Mary's-Milk, Souris, Red, Roseau, Rainy, St. Mary's, St. Clair, Detroit, Niagara, St. Lawrence, St. John, and St. Croix Rivers. Osoyoos Lake, Waterton Lakes, Lake of the Woods, Quetico-Boundary Waters, Lake St. Clair, Lake Champlain, and Lake Memphremagog. Plus, of course, four of the five Great Lakes—Superior, Huron, Erie, Ontario—that form North America's inland seas, the industrial epicenter of Canada and the United States from the mid-nineteenth to mid-twentieth centuries.[28] With over 20 percent of the world's available fresh surface water, the longest border shared by any two countries in the world is also the most fluid.[29]

With water then, Canada and the United States long faced disputes and mutual interests on a scale greater than most international waterways.[30] In the nineteenth century, the two countries struggled over shared water governance. Hard negotiations came to fruition with the Boundary Waters Treaty of 1909.[31] The treaty established a binational International Joint Commission (IJC) to resolve conflicts and facilitate mutual interests.[32] The treaty was also a diplomatic coup for Canada. Still under Mother Britain's wing in many regards—indeed, it was Britain that actually signed the treaty on Canada's behalf—Canada had gained parity with its more powerful neighbor.[33] Water now defined the Canadian-American relationship.

Out of the Boundary Waters Treaty of 1909, water law and policy would evolve differently than in regions that lay entirely within Canada or the United States. A wealth of water rather than a dearth of water often took center stage. Contemporary scholarship has rightfully critiqued both the Boundary Waters Treaty of 1909 and the International Joint Commission. Until the 1960s, the IJC promoted industrial development in border watersheds in spite of environmental and social harms. Other times, Canadian and American governments ignored or marginalized the IJC in favor of policies or projects that severely degraded water resources and hurt local communities. It's important not to sanitize an entire century of a major treaty.

But let's put on rose-colored glasses for a minute, and celebrate the vision of the treaty's article IV. Article IV states that "boundary waters and waters flowing across the boundary shall not be polluted on either side to the injury of health or property on the other."

Remember, this was 1909. There was no national movement to combat pollution, no areas of concern or public advisory councils. The modern environmental movement was fifty years in the future. The US Clean Water Act of 1972 and the US–Canada Water Quality Agreement of 1972 were more than sixty years in the future. For decades article IV clearly failed (just look at toxic pollution in the St. Lawrence River, the Detroit River, the St. Clair River, all border waters). And yet, in 1909, beneath that assertive stance of "shall not" lay an ambitious principle to anticipate and resolve future environmental conflicts.

Anticipation is the antithesis of the paradox of abundance. The paradox of abundance is reactionary: plunder first, response and recovery later. But along the Canada–US border, from the western Fraser River to the eastern Maritimes, and mid-continent at Lake of the Woods and the Great Lakes, the treaty provided a legal framework, and the IJC provided a forum, to anticipate, to study, and to negotiate alternative futures.[34] The IJC's pioneering research and policy formulations foreshadowed such future-oriented concepts as ecosystem management, the precautionary principle, and sustainability.

An imperative to shape their shared future: In short, this is why eight American states, two Canadian provinces, and two nations could come to an agreement on a legal, economic, and environmental matter as contentious as controlling water. The 2008 Great Lakes–St. Lawrence compact and agreement were a century in the making.

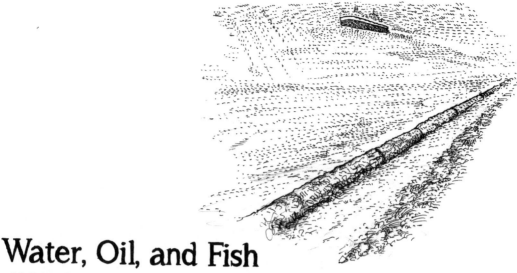

Water, Oil, and Fish

with Daniel Macfarlane

On Visualizing the Hidden: IIB and 6A

Consider the *where* of the photo below. A ragged patch of grass and gravel, a straight line of water slicing through flatland, a row of smokestacks across and parallel to the water whose silhouettes are reinforced by black belches drifting out of the frame on the right. On the far bank, a couple of trucks behind barbed-wire fence. Somewhere industrial.

Dominating the scene is a sign, "DANGER: ELECTRIC FISH BARRIER." This was the terminus of a field outing that began with a simple question, *Where is it?* Where was this new electric fish barrier, titled Barrier IIB? Barrier IIB was the US Army Corps of Engineers' latest reinforcement to a Great Lakes stronghold against bighead and silver Asian carp, those powerful nonnative fish that arrived from their journey up the Mississippi and encamped in the Chicago Sanitary and Ship Canal behind gates barring the way to Lake Michigan.

The location was Romeoville, in Will County, Illinois, about thirty miles southwest of downtown Chicago. While newspapers had announced the barrier, finding it wasn't easy, and once there, nothing barrier-like was visible. So the not-really-so-simple question of *where* morphed into another not-so-simple question of *what* was Barrier IIB.

Sometimes you have to be in a place—you have to take a look around—in order to ask questions, or to see something unexpected. Go back to the first photo and look past the danger sign. Explore the edges instead—for example, tree branches poking into the scene from the upper left, or the object below the branches, a metallic fragment arching across the water. *What . . .* is that?

That turned out to be Enbridge pipeline 6A. The second photo shows it close up. Line 6A is one segment of a binational crude oil trunkline system that begins in Fort McMurray, Alberta, heartland of the Canadian tar sands. Line 6A and IIB

Reading an unfamiliar waterscape. Photo by Lynne Heasley.

Unexpected intersections in the waterscape. Photo by Lynne Heasley.

intersected at the canal. Tar sands oil and Asian carp, jarring in their together-ness. Even more unexpected was their obscurity.

It may seem strange to call the pipeline obscure, since the arch over the canal is visible from a local overpass. In a way, it's hidden in plain sight. Boats and barges cruise underneath the structure, while cars go over a bridge that offers a distant view. But boaters and drivers must keep moving. They can't get off the boat to look around. They can't be *still* enough to study the place, to know it better. Perhaps the pipeline's operators count on a public moving rapidly past everyday sights. Richard White once wrote about water infrastructure that "boredom works for bureaucracies and corporations as smell works for a skunk. It keeps danger away. . . . The audience is asleep. The modern world is forged amidst our inattention."[1]

Line 6A's brief materialization aside, the landscape hosting 6A was most defi-nitely indistinct. This off-limits property concealed the actual location of 6A and IIB from passers-by. A tree-lined centennial trail lined one outer boundary of the property. But no trail map pointed inward to pipeline or fish barrier. There was a DANGER sign, after all. Looking for fish barriers meant turning away from the comforting habitual traffic of walkers, runners, and bicyclists. There were discreet entry points—a shallow part of a drainage ditch, a panel of knocked-over chain-link fence. A man walking two tiny dogs hopped the fence, then receded into the tire ruts and wiry gray-green stubble of an industrial barrens. Somewhere below ground, the pipeline continued unseen and unimpeded.

Obscurity matters. With water, oil, and fish, critical environmental histories of Chicago and the Great Lakes coexist out of sight, out of mind, and submerged. Pipeline and fish barrier represent the kinds of subterranean networks within which millions of people live unaware. The remainder of this chapter focuses on three intersecting Great Lakes infrastructures: the Chicago Diversion, Enbridge's Lakehead pipeline system, and the electric fish barrier system. Vast in the scale of their impact, precarious in the intricacy of their design, we try to make these concealed infrastructures visible by applying concepts for "seeing":

- Technological matrix of place. Together, diversion, pipeline, and barrier create what we call a "technological matrix of place": various infrastruc-tures that intersect at a discrete place. At the same time, each is part of a networked technological system.
- Disguised design. All three infrastructures are largely unseen despite their scale—underground, underwater, or off limits. They are emblematic of what Daniel Macfarlane has called "disguised design." These infrastructures are meant to be concealed.[2]
- Environmental risk. All three contain knowable risks to local communities and the larger Great Lakes region. Whether water, oil, or fish, in each sys-tem a resource flows through a conduit. The risk is that something suppos-edly confined and safe will escape its scripted boundaries.

We want readers to consider the repercussions of submerging both physical infrastructure and risk. We want readers to *see* how these infrastructures undergird not only the natural world and built environments of the Chicago area, but the natural world and built environments of the Great Lakes.

Matrix #1: The Chicago Diversion

The 1900 Chicago Sanitary and Ship Canal marked an audacious start to twentieth-century water engineering because it literally reversed the Chicago River. Today, historians and engineers see this canal system as a milestone in North American environmental history, and one of the century's top ten American public works projects.[3] The canal and its extended river network allowed Chicago to withdraw water from Lake Michigan, making it the first large-scale water diversion out of the Great Lakes. The "Chicago Diversion" was the volume of water the city could legally withdraw and send across the Great Lakes hydrological divide to the Mississippi River basin. From there, it would flow downriver to the Gulf of Mexico rather than from Lake Michigan to Lakes Huron, Erie, Ontario, and out the St. Lawrence River to the Atlantic.[4] This east-to-west, Chicago-to-Mississippi, orientation is typical of how people view the canal. By reorienting one's perspective from the canal toward the larger Great Lakes–St. Lawrence basin—that is, from west to east instead of east to west—we can see key but lesser-known relationships.

The canal itself was one example of a water engineering process that transformed the whole of the Great Lakes. Upgrades to the Welland Canal and locks at Sault Ste. Marie connected the lower and upper lakes. Dredged channels in the St. Clair and Detroit Rivers increased the capacity of Great Lakes harbors. Massive hydro-electric installations turned rivers into reservoirs. The St. Lawrence Seaway and Power Project was the most ambitious of these megaprojects; it opened the basin to transoceanic shipping. (The seaway, too, counts among the top ten public works of the century.)[5] These far-flung works formed an integrated system within which ships and their cargoes of Great Lakes resources circulated from the far western end of Lake Superior to the Atlantic Ocean. At 2,300 miles long, the Great Lakes–St. Lawrence waterway became the world's largest inland maritime waterway. It also became a transnational technological matrix.

The canal's place within a Great Lakes maritime system helps explain why a chronology of legal disputes now characterizes its history. At odds with Chicago, Canada and the American Great Lakes states had strenuously objected to the initial Chicago Diversion. They argued that it lowered Great Lakes water levels.[6] The diversion also became a thorn in US–Canadian diplomacy. Early on it factored into negotiations over the 1909 Boundary Waters Treaty. The treaty provided for joint management of Canadian-American boundary waters. Because of the Chicago Diversion, the final treaty did *not* include Lake Michigan.

Between 1912 and 1924 Canada filed six objections to the diversion. Later the diversion impacted bilateral talks for the joint St. Lawrence navigational seaway and hydropower development. Treaty negotiators for this Great Lakes Waterway Treaty of 1932 had included limitations on the Chicago Diversion. The US Congress voted down the treaty because of these provisions.[7]

US–Canada wrangling had technological and environmental consequences. In 1941 the countries signed an executive agreement that echoed the 1932 treaty. Congress rejected this agreement too. Canada reacted by building its own enormous Ogoki and Long Lac diversions *into* the Great Lakes, to compensate for the Chicago withdrawals.[8] The Ogoki–Long Lac megaprojects channelled water from James Bay and the Albany River into Lake Superior.[9]

In the 1950s, the Chicago Diversion reappeared during final treaty negotiations over the joint St. Lawrence Seaway and Power Project. The US Supreme Court had lowered the allowable volume of the diversion to 1,500 cfs (cubic feet per second), but on several occasions, Congress temporarily increased the volume, and it twice legislated a permanent increase. Canada formally objected to the legislation, asserting treaty violations. President Eisenhower promptly vetoed the legislation.[10]

Along with Canada, Great Lakes states also fought Chicago and Illinois over the Chicago Diversion. In 1967, the US Supreme Court established an average of 3,200 cfs limit for Chicago.[11] In the 1980s the Corps of Engineers studied an increase to 10,000 cfs. This was the maximum flow the canal could sustain. Later, Illinois formally, but unsuccessfully, requested the same increase. Chicago often exceeded its withdrawal limit, sometimes intentionally, sometimes by accident. More recently, the city has stayed within its legal limits.

The diversion itself has continued to weigh down Great Lakes policy and diplomacy. Illinois's refusal to consider any changes nearly sank a decade of interstate negotiations for a Great Lakes–St. Lawrence River Water Resources Basin Compact. The compact was a path-breaking framework for governing diversions out of the Great Lakes, as well as a model of water policy worldwide.[12]

The issue of water levels has also loomed large. Scientists in the early twentieth century could only hypothesize about fluctuating lake levels; the system-wide impacts of the Chicago Diversion were unclear.[13] But Canada and the Great Lakes states feared it could lower water levels by half a foot, and as far as Montreal.[14] Later research on Great Lakes hydrology revealed the complex ways in which the lakes are interconnected.[15] For example, engineering interventions in the Great Lakes–St. Lawrence basin *have* cumulatively lowered water levels, while natural forces have impacted the *scale* of the fluctuations.[16] The Chicago Diversion itself has slightly lowered water levels throughout the basin, while precipitation and evaporation have determined long-term fluctuations (including the record high water levels of recent years).

Private industry was likewise engaged in the pros and cons of the Chicago Diversion. Shipping and hydropower interests were especially hostile. Far to the east, along the St. Lawrence River, power providers had to raise their hydro stations and intake works to compensate for lower water levels. At Niagara Falls, lower water levels required remediation work on submerged weirs and control dams, and eventually the whole Niagara River cataract was reengineered.[17]

Industry faced feedback loops. Lower water levels reduced the weight of cargo that Great Lakes freighters could carry. Deeper wing dams and dredged bottoms became essential projects for industrial shipping. In the St. Clair and Detroit Rivers, engineers would periodically dredge and channelize the river. But those projects lowered water levels in Lakes Michigan and Huron.[18] Such iterative actions and reactions, causes and effects, make visible an underwater infrastructure matrix in which a technological "butterfly" flapping its wings in one place had distant hydrological, ecological, and economic impacts elsewhere in the system.

For one-hundred-plus years, the Chicago Diversion was a sore infecting interstate politics and international diplomacy. Impacts below the surface were just as important. The brute force of the diversion altered water levels and forged new hydrological networks. By lowering lake levels, the diversion shaped control works: locks, dams, weirs, channels. Each looked like a single infrastructure project initiated in response to local conditions. Together, though, they formed a highly engineered Great Lakes system. Also, below the surface were disrupted lake bed and river bottom morphologies and aquatic ecosystems. The Chicago Diversion transformed the waterscapes and water life beneath.

Matrix #2: The Enbridge Lakehead Pipeline System

Like water flows from the Chicago Diversion, oil flows through the Chicago region are mostly invisible. At the canal near the Chicago suburb of Romeoville, Enbridge pipeline 6A makes a rare aboveground appearance: a dramatic aerial line featuring a double wishbone arrangement of two thirty-four-inch pipes. The striking arch reaches a height of ninety-three feet. But even this singularity conceals what runs beneath, underground or underwater, a continent-scale network of fossil fuel pipelines.

Enbridge alone has seventeen thousand miles of pipeline in the United States. The Lakehead System accounts for about 1.7 million barrels per day, or 13 percent of US petroleum imports. Enbridge does not disclose how much oil running through its Lakehead System comes from Alberta's huge tar sands deposits. In Chicago, the Lakehead System is one of the most important infrastructure matrices that people don't see and know little about.

The first long-distance pipelines in the United States ran near and partially within the Great Lakes basin, especially the southern shore of Lake Erie.[19] The Petrolia–Sarnia region of Ontario, Canada, bordering Michigan, was ground zero

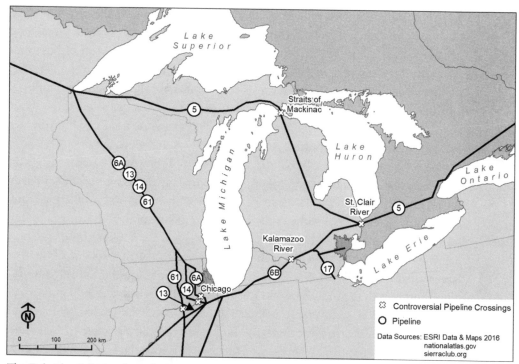

The Enbridge Lakehead pipeline system traversing the Great Lakes basin.
Map by Jason Glatz, Lynne Heasley, and Daniel Macfarlane.

for Canadian production; there oil interests in the era of John D. Rockefeller's
Standard Oil built Canada's earliest pipelines. By the end of the nineteenth century, Ontario and Michigan had underwater gas pipelines at the bottom of the
Great Lakes connecting channels like the St. Clair River. Twenty-eight pipelines
now run through the St. Clair.[20] One pair dating to 1918 was long forgotten until
2016, when new owners quietly applied for a permit to reactivate the antique
pipes for transporting liquid hydrocarbons, including crude oil. St. Clair communities learned about the pipes *and* the permit application only in the last days of
the public comment period.[21] Under sudden public scrutiny, the application was
withdrawn.

Pipeline infrastructure scaled up after World War II. In 1950, Canadian energy company Inter-Provincial Pipelines—now Enbridge—opened an oil pipeline from Edmonton, Alberta, to Superior, Wisconsin. This was the first leg of
a network that includes Enbridge's Lakehead System.[22] From Fort McMurray in
Alberta, Enbridge's main line runs to Superior, Wisconsin, at the far western tip
of Lake Superior. At Superior the system forks into line 5 and line 6A (built in 1953
and 1960 respectively).

Line 6A is Enbridge's largest oil artery supplying its US market. From Superior
line 6A goes southeast through Wisconsin and Illinois. Near the Chicago Sanitary

and Ship Canal, 6A runs underground, then soars briefly over the canal. In April 2010 at Romeoville, municipal employees, rather than Enbridge itself, discovered a major spill from line 6A. All told, six thousand barrels spilled.[23] Authorities evacuated five hundred people from area businesses when the oil spread to the retention pond of a nearby wastewater treatment plant. Enbridge closed three miles of pipeline in order to locate the leak. The closure led to higher gas prices throughout the Midwest.[24] Tom Kloza, publisher and chief oil analyst at Oil Price Information Service, said it was like "shutting down the lifeblood that feeds the organs."[25]

Line 6A ends southwest of the canal, at Griffith. From there its sister line, 6B, continues through Indiana and Michigan to Port Huron, and under the St. Clair River to its terminus in Sarnia. Many people in Michigan know Line 6B even if they don't know its name. Line 6B was the infamous pipeline of the July 2010 Enbridge oil spill into a tributary of the Kalamazoo River—over 1.2 million gallons (20,000 barrels) of tar sands diluted bitumen, or heavy crude oil. This was one of the two largest inland oil spills in US history (both from Enbridge lines).[26] Enbridge's combined costs for the 2010 Romeo and Kalamazoo spills exceeded $1 billion.[27]

For its part, line 5 traverses northern Wisconsin and the Upper Peninsula of Michigan, turns south, runs underwater through the Straits of Mackinac, and continues to Port Huron, where Lake Huron drains into the St. Clair River. Like 6B, line 5 terminates in Sarnia. Sarnia serves as a transfer point for oil heading inland, but the city's so-called Chemical Valley also holds 40 percent of Canada's petro-chemical industry.

Line 5 has become infamous in its own right. Only in the last decade did Michigan and Ontario communities learn that a major pipeline went through the Straits of Mackinac, a natural channel that connects Lakes Michigan and Huron, making them hydrologically, if not in name, a single lake and the largest freshwater lake in the world by surface area.

With line 5, a perfect storm of conditions is poised for disaster. The straits' "strong and complex" currents buffet and stress the pipelines and in case of a spill would move oil quickly through the lakes.[28] There is no way to slow or remediate a spill if it occurs in icy winter conditions, which last for many months of the year. Sediment covers parts of the pipeline, so its condition is unknown in those places. In other exposed places film footage has shown broken supports and corrosion, unsurprising given that the pipeline is nearly seventy years old.[29] "If you were to pick the worst possible place for an oil spill in the Great Lakes, this would be it," said hydrodynamic modeler David Schwab.[30] Yet each day, two hundred feet underwater, line 5 carries 20 million gallons of synthetic crude oil across 4.5 miles of straits. Michigan politicians and policymakers face tremendous public pressure to force Enbridge to shut off the oil through line 5.[31]

Legally, waters like the Kalamazoo and St. Clair Rivers or the Chicago Sanitary and Ship Canal are held in "public trust." Unlike land, no private water owner can buy acreage of river or lake bottom. Parties may only negotiate a conditional right-of-way with the state. With respect to line 5 in the Straits of Mackinac, perhaps the most shocking discovery was how little anyone knew, including who was responsible for its safety. It turned out that, despite the Kalamazoo and Romeo spills from the same Lakehead System, the State of Michigan had not revisited its original 1953 easement for the Enbridge pipeline. Nor had the governing federal agency, the Pipeline and Hazardous Materials Safety Administration (PHMSA), reviewed line 5. Until underwater footage revealed line 5 to an upset public, the PHMSA had been nearly as disguised as the pipeline under its watch.

The canal, straits, and rivers are among the thousands of places where oil crosses above or beneath water. Each is a submerged technological matrix, a vulnerable hybrid of engineered and natural hydrological systems. Collectively they form an interconnected network in which leaks and spills are the norm. Since 1999 North American Enbridge pipelines have lost 6.8 million gallons from over eight hundred leaks and spills.[32] Industry-wide, spills occur daily.

Transnational pipeline networks have been called a "dissociating technology." They lie out of sight. They don't need human hands to operate until oil escapes. And they are geographically distant from the original source of the oil.[33] Their disguised design means that communities, and even the pipeline operator, won't immediately know of leaks and spills, so time lags before discovery are also the norm.[34]

By design, a continent-scale pipeline system hides, delays word of, and eventually externalizes the environmental and economic fallouts of oil production to local landowners and communities, and to local landscapes and waterscapes. The system even threatens immense waters like the Gulf of Mexico or, forebodingly, the Great Lakes. Making these infrastructures visible is a precondition for safer policy.

Matrix #3: Electric Fish Barriers

In 1900 Chicago completed its eight-year quest to flush the city's excrement west, to Lockport and the Des Plaines River and finally to the Mississippi. City leaders celebrated more than Chicago's new stature as a model of modern wastewater treatment. The Chicago Sanitary and Ship Canal symbolized clean, safe water. Hence city leaders could celebrate the blurring and eventual erasure of Chicago's earlier image as a cesspool—a literal cess-river system of human and animal waste oozing through streets, clogging sewers, thickening and chunking the river, rendering the South Fork of the South Branch of the Chicago River bloody and bubbling with the methane and hydrogen sulfide gasses of decomposing hog and cow entrails from the meatpacking district. The South Fork of the South Branch had earned its moniker, "Bubbly Creek." But the new canal flushed the city clean. The canal also scrubbed away the public's imagination of Chicago as a place of disgust and degradation.

The Chicago Sanitary and Ship Canal is back in the public imagination. This time, though, the canal itself embodies a potential catastrophe: biological degradation, or "biological pollution" of the Great Lakes. The context is this: The unification of the Great Lakes and Mississippi basins created new flows of nuisance aquatic species (NAS, in the technical literature). Zebra and quagga mussels and round gobies crossed the Atlantic Ocean to the Great Lakes via the St. Lawrence Seaway, and then migrated to the Mississippi basin via the Chicago Sanitary and Ship Canal. These species fanned southward along the Mississippi River and its tributaries to the Gulf Coast.

Yet the canal was not a one-way-only biological highway. Notre Dame scientist David Lodge called the route "a two-way highway to environmental and economic havoc."[35] Lodge was giving congressional testimony about the biogeography of silver and bighead carp, *Hypophthalmichthys molitrix* and *Hypophthalmichthys nobilis*, two of the species colloquially called Asian carp.

From their 1972–1973 introduction to the American South for use as algae cleaners in catfish farms, Asian carp escaped into the wild. They migrated up and down the Mississippi River system, decimating aquatic food webs along the way—"like a school of aquatic bullies," according to an Illinois Department of Natural Resources article.[36] In many parts of the Mississippi basin, silver and bighead carp constitute up to 95 percent of the total aquatic biomass. They reached Illinois rivers in the early 1990s, and in 2002, scientists saw them in the upper Illinois River, which is the conduit between the Chicago Sanitary and Ship Canal and the Mississippi River.

At the same 2010 hearing where Lodge testified, committee chairman and US representative James L. Oberstar likened Asian carp, "this treacherous, dangerous species that we cannot allow into the lakes," to sinister aquatic wolves: "It reminds me of an image in the language of my ancestors, the Slovenes," Oberstar said, "we just think about the wolf, and it is at our doors. And that is what the carp is; it is at our doors."[37]

The Oberstar hearing was focused on plans for the US Army Corps of Engineers Electrical Dispersal Barrier System. Technology had assumed a principal role in addressing the continental problem of invasive species like the Asian carp. In Chicago, electric fish barriers now form another disguised infrastructure, a matrix at once powerful and vulnerable—with both its power and its vulnerabilities determining the fate of Great Lakes ecosystems.

Today the system consists of barriers placed at intervals in the canal near Romeoville, Illinois. Planning for the barriers got underway in 1996, when Congress appropriated $750,000 for a "Dispersal Barrier Demonstration" as part of the National Invasive Species Act of 1996.[38] The Army Corps of Engineers activated demonstration barrier I in 2002 to test the technology, IIA in 2009, and IIB in 2011, following the Oberstar hearing. On shore, calibrated generating stations run and monitor the barriers. In 2017, the Trump administration put on hold an updated corps' study, which recommended another electric barrier farther south,

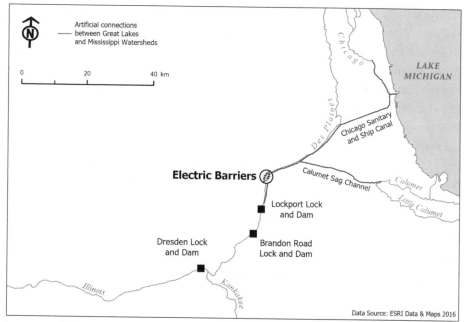

Location of the electric dispersal barrier system along the Chicago Sanitary and Ship Canal. Map by Jason Glatz, Lynne Heasley, and Daniel Macfarlane.

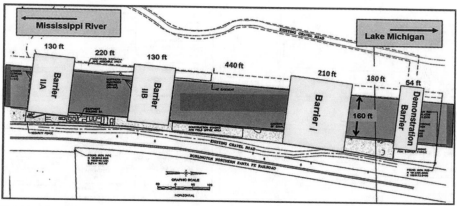

Schematic of Barriers I, IIA, and IIB in the canal. Source: US Army Corps of Engineers.

and farther away from Lake Michigan, at the Brandon Road Lock and Dam, a Des Plaines River choke point for carp; however, these plans are again underway.

Smith-Root Inc. was the corps' sole contractor to design and build the dispersal barrier system. The company is an example of historical happenstance. In 1964, Dave Smith and Lee Root designed an electric fisher for University of Washington fish biologists. Their invention made fisheries research more efficient. Smith and Root had stumbled into an unexpected market. "They got drug into the electro-fishing market, not by design, not by creativity; it just happened," said Smith-Root president and CEO Jeff Smith.[39]

Smith-Root soon expanded into other electric devices, including complex systems for redirecting or containing fish traffic. This was a technology-science collaboration. Smith-Root worked with fisheries and other aquatic scientists nationwide, and conversely, scientists assumed a proactive role in emerging technological approaches to environmental problem solving. Electric dispersal systems like that in the Chicago Sanitary and Ship Canal marked the culmination of this codependence.

From 1988 through 2016, Smith-Root filed six patents for some version of electric dispersal fish barrier systems. Barrier I used Smith-Root's original 1988 design for a "fish repelling apparatus using a plurality of series connected pulse generators to produce an optimized electric field."[40] A sample patent diagram below shows the layout, crudely stated here, of electrodes along metal strips at the bottom of the canal supported by electric generation on land, and with cables connecting the electrodes to the generators. The little fish icons represent the carp. In subsequent patents, Smith-Root expanded their barrier systems.

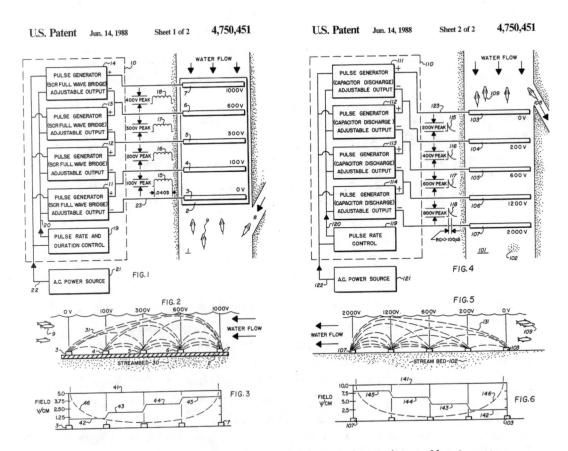

Illustrations from US Patent 4,750,451, a Smith-Root electric dispersal barrier system.

The fish icons of Smith-Root's patent diagrams are easy to overlook, yet they might be the most telling part. Buried within the complex energy matrix of generators, circuits, transformers, and electrodes were fish physiology and behavior. Controlling the fish depended on the precise application of current running through water. This in turn depended on intimate knowledge of how electricity passes through a moving animal's body and shapes its behavior. For instance, in Smith and colleagues' 2005 patent, "Electric fish barrier for water intakes at various depths," the team detailed such electricity-fish interfaces: "Because a fish has salts and electrolytes within its body . . . a fish's body acts as a 'voltage divider' when swimming through fresh water."[41] But precision could never be wholly achieved, because these were real-world uncontrolled experiments in which two intersecting variables—animals and electricity—were not fully known or predictable.

Smith-Root and their university partners raced to study carp behavior around electricity. They hoped to adapt the system to newly observed behavior before the fish found their way through the barriers and into Lake Michigan. Each update of the electric dispersal barrier system became a new field trial. The most alarming carp behavior involved differences between large and small animals. It turned out that juvenile fish could withstand a higher level of pain before turning away from the barrier, even when the acute pain made them swim abnormally. The initial electric pulses had emitted too low a voltage to repel juvenile carp. Voltage adjustments were made. Juveniles could also slip through bedrock cracks in the waterway's surrounding landscape. Regional flooding thereby became high-risk events. Strategic fencing in vulnerable locations followed.

Add in carp tenacity: biologically they were impelled to move upstream, so over and over they probed and tested the barriers, finding less discomfort closer to the surface and less discomfort toward the canal edges. Acoustic bubbles or bubble barriers near canal walls might address this particular vulnerability.

Barges and other vessels on the canal presented a risk, altering the direction of the electric current so that fish might draft off them all the way through a barrier.

And then there was safety: Could the electric current escape the canal and hurt people? Steel "parasitic structures" across the canal dealt with this concern.

The corps entered all these known risk variables into a risk model.[42] But the upshot was that carp would adjust, would search for ways to get through, and their behavior never went entirely according to plans.

The history of the barrier system is inseparable from the history of late twentieth-century fisheries research. A codependence of science and technology formed under the intense pressure of fast-spreading invasive species whose ecological and economic costs were enormous. But this same interdependence may have foreclosed or delayed other options. The safer, longer-term solution would be to re-separate the Mississippi and Great Lakes basins. Hence many of the same Great Lakes stakeholders who tried to block the canal's opening more than a hundred years ago are now trying to close the canal. So far, however, an alliance of city leaders, Illinois state leaders, heavy industry (e.g., oil), Mississippi River tow and barge companies, and Chicago tourist operators have prevailed over seven Great Lakes states, two Canadian provinces, a $7 billion Great Lakes fishing industry, and the nearly unanimous ecological anxiety of aquatic ecologists and fish biologists.

Where do matters stand, or drift, as it were? Anglers have caught several types of Asian carp in the Great Lakes, but not yet silver or bighead species. Scientists have detected environmental DNA (eDNA) of silver carp in the Chicago River and Great Lakes, though eDNA alone does not mean there are reproducing populations. In 2010 the Illinois Department of Natural Resources reported a nineteen-pound bighead carp in Lake Calumet, past the electric barriers and six miles from Lake Michigan. In June 2017, a commercial fisherman caught an eight-pound silver carp past the barriers and nine miles from Lake Michigan.[43] Such findings mean that carp are finding ways to Lake Michigan, suggesting it's just a matter of time until the fish establish a foothold in the lake.

In the case of oil pipelines, the submerged infrastructure facilitates the dispersal of the environmental risk. In the case of electric fish barriers, the submerged infrastructure is the complex but precarious matrix designed to *contain* the spread of the risk. Celebrated for cleaning up Chicago at the dawn of the twentieth century, the canal is now a threatening conduit for biological pollution. While the Chicago Diversion influences Great Lakes water levels far to the east, and oil pipelines form a techno-geographical relationship with the west and north, invasive carp trace a route from the southern United States through the metropolis and into Lake Michigan. Chicago is connected to most of the continent by concealed technological matrices that intersect at the Sanitary and Ship Canal.

Disguised by Design

In *Nature's Metropolis*, William Cronon's classic study of nineteenth-century Chicago, the city was the gateway through which railways and shipping lanes funneled and dispersed resources cum commodities.[44] The water, oil, and fish of this chapter have a similar metropolitan–hinterland relationship. Maps of contemporary pipelines converging on and then diverging from the Chicagoland region echo nineteenth-century railway maps. But instead of Cronon's pork and lumber, Chicago is now a hub moving fossil fuels and fish across the continent.

Cronon also posited a "second nature," a city built from nature transformed and commodified. Since *Nature's Metropolis*, a robust envirotech scholarship has approached landscapes and waterscapes as intertwined hybrids of nature and technologies.[45] Thus we arrive at ways to understand a twenty-first-century Chicago, where boundaries between the natural world and technology are porous; where water, oil, and fish were built into the infrastructures of canal, pipeline, and barrier; and where the infrastructure itself became part of a multidisciplinary scientific enterprise. In present-day Chicago, an electric fish barrier is a technology whose success and vulnerability hinge on the natural phenomena of electricity moving through water and through fish.

Built environment is a common academic term for urban landscapes and waterscapes. The Chicago Sanitary and Ship Canal is also a system of "built ecologies." In the canal, natural elements and processes are the dominant aspects of what is nonetheless a human-constructed system. We suggest a unique category of built ecologies and technological megaprojects that are hidden (or at least hidden in plain sight).[46] This category stands in contrast to engineering projects celebrated for their public visibility, such as epic hydropower dams proudly displayed by the state and admired by awestruck masses.[47] By design, water diversions, buried pipelines, and fish barriers are *unseen*.

Disguised infrastructures have virtues. They are conveniently out of people's way, ingenious in their subtlety. They may even provide peace of mind, because sometimes we don't *want* to see. Disguising infrastructure, however, hides intentional environmental and economic risk-taking. The problem is that those taking risks with hydrology, oil spills, and aquatic ecosystems will not bear the brunt of disasters. Any economic or ecological fallout will be at once highly localized and widely regionalized. The changed hydrology of a reversed Chicago River reverberated throughout the Great Lakes–St. Lawrence waterway. The Chicago Sanitary and Ship Canal became a two-way ecological corridor across two of North America's largest basins, and between North America and Eurasia. Oil pipelines linked Chicago to the Great Lakes in new ways, and the Great Lakes to North America. These disguised infrastructures are implicated in, necessary for, and reliant on continental and global networks.

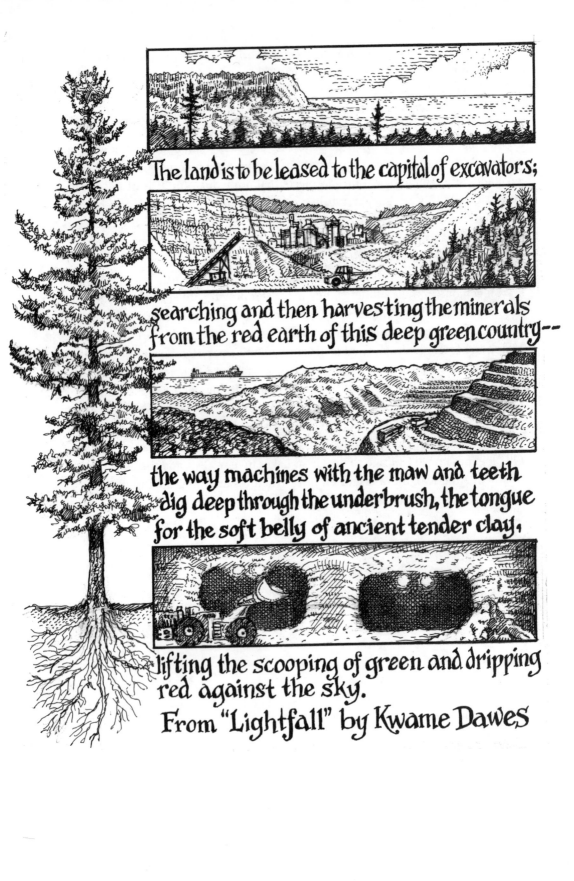

The land is to be leased to the capital of excavators;

searching and then harvesting the minerals
from the red earth of this deep green country--

the way machines with the maw and teeth
dig deep through the underbrush, the tongue
for the soft belly of ancient tender clay,

lifting the scooping of green and dripping
red against the sky.
From "Lightfall" by Kwame Dawes

Salt Mines and Iron Ranges (An Extraction Index)

Note to Readers

This is an experimental piece—a thought exercise—on resource extraction. I'm adapting the form of the Harper's Index. In 1984 Michael Pollan helped launch the Harper's Index. He was senior editor and soon after executive editor of *Harper's*, and not yet one of the foremost writers on our industrial food system. The Index is a list of forty entries, each ending with a number. I haven't mastered the form like the *Harper's* research team (they're the professionals). But the form is suited to the immensity of extraction in the Great Lakes and on Earth. I try my hand with trees, salt, iron, and sand—stacking, grouping, sequencing entries so that each index touches on:

Immensity
Stories and histories
Great Lakes<–>global connections
Brutality and rejection thereof
Sustainability and restoration
Wonder

Each entry within an index is a story lede but not the story, a first breadcrumb down a different trail. Each index is a conglomerate, changeable but not interchangeable. Your entries would be different than mine. Your index would form a different conglomerate.

Finally, numbers have curious effects: A person suspicious of someone's emotion on a topic might gain confidence with a number. A person sure that numbers lie, deceive, and obscure might argue with the numbers, or turn away. Numbers can overwhelm, shell-shock. Worst of all, numbers can numb, as being force-fed a tranquilizer. But when numbers involve a place you love—like the Great Lakes, or some other place in the world—hopefully you'll be moved to intervene in your own way.

Trees

Age (in years) of Pando, a quaking aspen cluster and oldest clonal organism on Earth : 80,000

Age of Methuselah, a bristlecone pine and the oldest tree on Earth : 4,600

Height of the tallest white pine "monarchs" of the Great Lakes : 200'

Daily tons of water a broad leaf tree draws from the soil up its trunk to the tips of its leaves : 1

Years before present when white pine migrated to the Great Lakes region : 10,500–7,200

Years CE of the pine cutover in Michigan, Wisconsin, Minnesota (the Lake States) : 1860–1920

Number of horses the Diamond Match Company used in Michigan's largest pine cut : 1,200

Football-field lengths of Frederick Weyerhaeuser and Edward Hine's Chicago lumber dock : ~14

Dimensions of one board foot of lumber : 1 ft. × 1 ft. × 1 in.

Approximate board feet of white pine cut from the Lake States during the cutover : 270,000,000,000

Ratio of trees cut and sold to trees wasted or burned in the cutover : 2:1

Percent of the five deadliest wildfires in US history that occurred in the Lake States : 100

Estimated human deaths in the Peshtigo, Wisconsin, "holocaust" firestorm of Oct. 8, 1871 : 1,500–2,200

Estimated tree deaths in the Peshtigo firestorm : 2,000,000,000

Estimated animal deaths in Australia's wildfires of 2019–2020 : 3,000,000,000

Approximate board feet of wood in both a Ford Model T and a mature sugar maple tree : 250

Acres of woodland Henry Ford bought in the Upper Peninsula : 430,000

Horsepower of the first one-man chainsaw "the Beaver" in 1944 : 1.25

Year Michigan created Porcupine Mountains Wilderness State Park over industry objections : 1944

Miles from Earth satellites show the exceptional forest of the Menominee Indian Tribe of Wisconsin, including old-growth white pine : 435

Number of trees and shrubs that Great Lakes Anishinaabe used for medicinal purposes : 90

Equivalent acres of Michigan forest needed to build a modern 2000-square-foot home : 3–7

Number of Bishnoi beheaded in 1730 as they hugged their Khejri trees to stop a Maharaja's troops from cutting the trees for a new palace : 363

Number of environmental (mostly forest) defenders murdered between 2012–2017 : 1,558

Number of rubber-tree seeds that Henry Wickham stole from the Brazilian Amazon in 1876 for plantations in India and Asia : 70,000

Year a cattle rancher murdered Brazilian rubber tapper/rainforest activist Chico Mendes : 1988

Percent of current deforestation in the Amazon region caused by cattle ranching : 80

Rank of deforestation as a cause of global climate change : 2

Tons of carbon stored by the Amazon rainforest : 180,000,000,000

US ranking of the Lake States in density of carbon sequestration : 1

Annual pounds of carbon stored by one acre of mature oak forest : 30,000

Number of passenger pigeons that could roost and nest in one mature white oak : 500

Number of butterfly and moth species sustained by one mature oak : >550

Economic worth of ecosystem services provided by trees in urban areas : $2,400,000,000

Year the first splinter-free toilet paper came on the market : 1930

Number of toilet paper rolls Americans use in a year : 36,500,000,000

Percent decrease in number of trees since the beginning of human civilization : 46

Year Englishman John Evelyn said, "Men seldom plant trees till they begin to be wise" : 1664

Record number of trees Ethiopians planted in twelve hours on July 29, 2019 : 353,633,660

Acreage of Boreal Forest in Canada and Alaska that remains mostly intact : 1,200,000,000[1]

Salt

Year BCE Alexander the Great's horses licked salty rocks near today's Khewra Salt Mine in Pakistan (second-largest in the world) : 326

Number of natural salt licks wildlife used on the Ojibwe Sand Ridge Trail near Lake Huron : 3

Number of cows in May Swenson's 1955 "Eclogue," who lick her naked prone body, believing she was "a slab of salt discovered on this miraculous day in their hilly meadow!" : 10–12

Pounds of salt a herd of 100 pastured cattle needs weekly : ~55

Average IQ rise from Morton's 1924 iodized table salt, which reduced iodine deficiency : 3.5

Percent of the human body consisting of sodium chloride (salt) : 0.9

Years BP Michigan basin salt beds formed as an ancient sea evaporated : 420,000,000

Year geologist R. C. Allen declared Michigan brines and salts to be inexhaustible : 1914

Estimated tons of salt below the Lower Peninsula of Michigan : 30,000,000,000,000,000,000,000

Per bushel bounty that Michigan offered in 1859 to incentivize industrial salt extraction : 10¢

Fold increase of barrels of salt Michigan produced from 1860–1887 : 500

Miles of roads in the Detroit Salt Mine, 1200 ft. beneath the city : 100

Number of major salt mines under Lake Erie : 2

Global rank by size of the Goderich salt mine under Lake Huron : 1

Year BCE that the Chinese emperor Hsia Yu levied a salt tax : 2200

Year Mahatma Gandhi led a Salt March to the sea in protest of British colonial salt taxes : 1930

Number of references to salt in the Hebrew Bible : 35

Camels in the largest Saharan caravans to ply the Taoudenni salt mine and Timbuktu : 10,000

Annual tourists at the Chapel of the Blessed Kings, in Poland's Wieliczka Salt Mine : 1,000,000

Continuous days in 1901 of a seemingly providential oil geyser from Spindletop salt dome in Texas : 9

Amount of lithium in Bolivia's Salar de Uyuni salt flat as a fraction of global demand : ½

Year BCE Chinese hydraulic engineer Li Bing sunk the first recorded brine well : 252

Year Herbert Henry Dow leased a Michigan brine well to make chlorine and caustic soda : 1890

Equation for the chloralkali process of salt electrolysis, yielding chlorine, caustic soda, and hydrogen : $2NaCl + 2H_2O \longrightarrow 2NaOH + H_2 + Cl_2$

Industrial chloralkali and other chemical manufacturing as a fraction of global salt demand : ~⅔

Number of known uses for salt : 14,000

Year New Hampshire became the first state to use salt on icy roads : 1938

Year Michigan investigated groundwater contamination in Manistee (dubbed "Salt City of the Inland Seas") by Morton and other salt and chemical companies : 1968

Average tons of road salt now used during a US winter : 15,000,000

Annual costs to repair infrastructure damaged by salt : $5,000,000,000

Annual economic losses of salt-damaged cars : $3,000,000,000

Estimated percent of private wells in New York State with unsafe salt levels from road salt : 24

Estimated number of freshwater lakes in North America threatened by increased salinity : 7,770

Percent increase of salt content in rivers in the past twenty years : 100

Milligrams per liter of chloride in water that can cause sex reversals in frogs : 100–1000

Percent of road salt reduction possible by switching to liquid brine and best practices : 75

Quintillion kilograms of salt in the world's oceans : 50

Percent salinity of the Antarctic's Don Juan Pond, the world's saltiest water body : 44

First demonstration of reverse osmosis membranes for desalination of saltwater : 1959

Number of desalination plants worldwide to convert saltwater to freshwater : 17,000[2]

Iron

Age in years of the world's oldest smelted iron artifacts : 4,000

Tons of iron in the world's *available* ore resources : >200,000,000,000

Annual tons of iron ore extracted worldwide : 2,000,000,000

Number of elements that can be combined with pig iron to produce different steel alloys : 20

Parts per millimeter of the element vanadium in the "wootz" steel of famed Damascus swords : >40

Weight of a Ford Model T made with vanadium steel compared to other cars of the time : ½

Number of steel alloys made today : >3,500

Percent of iron ore used in the US steel industry supplied by Minnesota and Michigan mines : 98

Cubic yards of Lake Michigan coastal sand dunes removed in 1906 to build US Steel's integrated steel mill complex Gary Works, "industrial wonder of the world" : 12,000,000

Percent decline of the Gary Works workforce from 1970 to 2015 : ~84

Decade when Minnesota's Mesabi Iron Range ran out of high-grade iron ore to mine : 1950s

Decades that University of Minnesota engineer E. W. Davis worked to overcome technological, economic, and political barriers to low-grade iron ore (taconite) processing : 4

Tons of taconite on ore freighter Edmund Fitzgerald when it sank in Lake Superior : 26,116

Tons of taconite tailings E. W. Davis Works dumped daily in Lake Superior, 1956–1980 : 67,000

Asbestos fibers per liter in Duluth drinking water from E. W. Davis Works tailings : 100,000,000

Rank of Brazilian iron giant Vale *before* its Brumadinho tailings dam collapsed, killing 300 : 1

Assets held by institutional investors seeking corporate disclosure of tailings dam risks : $12,000,000,000,000

Year Krueger Brewing Co. sold the first beer packaged in steel cans : 1935

Number of epoxy coatings used inside beverage and food cans : 15,000

Average percent of a can's epoxy coating consisting of hormone disruptor bisphenol-A : 80

Pounds a 1970s American car could lose annually because of rust : ~10

Rank of rust among threats to United States Navy ships and vessels : 1

Percent of the Statue of Liberty's rusted skeleton replaced by stainless steel during restoration : 100

Year filmmaker Ken Burns interviewed writer James Baldwin for *The Statue of Liberty*, and Baldwin said, "For a Black American, for a Black inhabitant of this country, the Statue of Liberty is simply a very bitter joke, meaning nothing to us" : 1985

Estimated number of enslaved people who worked in the antebellum South's iron industry : >10,000

Number of northern Black iron and steel workers by 1920 : 125,000

Percent of the Mesabi Iron Range's mining workforce that was foreign born by 1920 : 85–90

Major strikes on the Iron Range lead by Finnish miners : 1907 and 1916

Number of workers in the "Great Steel Strike of 1919," after US Steel rejected outside unions : 367,000

Number of Black workers, barred from unions and better mill jobs, who crossed picket lines, causing the 1919 strike to fail : 30,000–40,000

Rank of iron and steel in industrial energy use and CO_2 emissions : 2

Reduction in CO_2 emissions with scrap-steel mini-mills compared to traditional mills : ¾

Fraction of end-of-life commercial ships beached in South Asia for dangerous, toxic "shipbreaking" into scrap metal: 2/3

Pounds of hexavalent chromium Gary Works–Midwest Plant spilled in Lake Michigan in 2017 : 902

Fish killed by steelmaker ArcelorMittal's cyanide spill at Burns Harbor, Indiana, in 2019 : 3,000

Number of Iron Range lakes and streams whose sulfate pollution can kill native wild rice : 60

Number of steel rivets in the Mackinac Bridge, North America's longest suspension bridge : 4,851,700

Height of the steel-concrete Burj Khalifa skyscraper in Dubai : 2,723'

Metric tons of global steel output from 1850 to 2015 : 51,000,000,000

Year Walt Whitman's *Passage to India* hailed the technological sublime of global industry—"The earth to be spann'd, connected by network. . . . The lands to be welded together" : 1871[3]

Sand

Number of grains of sand on earth in quintillions : 7.5×10^{18}

Number of grains of sand we can hold in one hand: 10,000

Sand mining as a percent of all mineral extraction on earth : 85

Rank of sand in global consumption of natural resources : 2

Rank of water in global consumption of natural resources : 1

Ratio of sand mined from China's Pyong Lake to the three largest US sand mines combined : 3:1

Percent expansion of Singapore's land area using Indonesian sand : 20

Number of Indonesian islands sand-mined out of existence : 24

Average percent of quartz (silicon dioxide) in Great-Lakes-coast sand : 90

Average percent of quartz in "frac" sand for hydraulic fracturing of shale oil and gas : >99

In Wisconsin's "Northern White" sand : 99.8

Required for use in solar panels : 99.999999

For use in computer chips : 99.99999999999

Position of the United States in the global silica sand market : 1

Position of Wisconsin in the US silica sand market : 1

Number of US construction workers exposed to respirable crystalline silica dust : 2,000,000

New sand mines and processing plants in western Wisconsin from 2010–2013 : 105

Percent of frac sand mined in the Great Lakes region as of 2014 : 70

Tons of sand needed to hydraulic fracture (frac) one shale oil well : 10,000

Number of wells in Michigan that used fracking : 12,000

Barrels per day increase in US oil production from the fracking boom, 2011–2018 : 5,900,000

Shale oil workers laid off in 2015 because of overproduction and plummeting prices : 55,000

Decline of oil and gas market value since 2016 : $400,000,000,000

Year early fracking pioneer Chesapeake Energy entered Chapter 11 bankruptcy : 2020

Year Chesapeake Energy CEO Aubrey McClendon of Oklahoma bought 402 acres of Saugatuck Dunes for a resort and marina : 2006

Year the Saugatuck Dunes Coastal Alliance formed to save Saugatuck Dunes : 2006

Estimated number of interdunal wetlands in one small section of Saugatuck Dunes : 123

Number of invertebrate taxa identified so far in these wetlands : 93

Age of pioneer ecologist H. C. Cowles when he published his 1899 study of plant succession on Lake Michigan's Indiana Dunes (now Indiana Dunes National Park) : 30

Global rank of Lake Michigan in number of freshwater coastal dunes : 1

Age of Lake Michigan's oldest coastal dunes : 3000–5000

Former height in feet of the landmark Pigeon Hill sand dune in Muskegon, Michigan : 217

Year Muskegon tried to save Pigeon Hill from sand mining by new owner Nugent Sand : 1925

Decades it took for sand mining to raze Pigeon Hill : 3

Year the state of Michigan passed the Sand Dune Protection and Management Act : 1976

Tons of sand mined from Lake Michigan dunes after the act, between 1978 and 1998 : 46,533,730

Percent of sand mined from Lake Michigan dunes used for industrial foundries : 95

Percent of sand used in cement mortars that could be replaced by incinerator ash : 40

Active coastal sand-dune mining sites on Lake Michigan as of 2019 : 9

Year *Good Morning America* viewers voted Sleeping Bear Dunes most beautiful place in America : 2011[4]

The Paradox of Abundance

Call it a sub-genre of historical writing about the natural world. The environmental film noir. Or maybe the ecological horror flick. The story might unfold anywhere, and at many times. Here stands a forest housing and fueling and feeding a people, an immense forest stretching across the latitude of a country, up the mountain slopes of a civilization, holding back the desert sands of a continent, or the tides of an ocean, imprisoning the carbon of a planet—acacia, palm, mangrove, mahogany, black spruce, white pine, Lebanon cedar, English oak.

Or there, below the surface, lies layer upon layer—seams—of magical rocks. Use heat, hammer or alchemy to unlock the metals within these rocks, and you unlock such power to control and transform our material world that Roman poet Ovid and Enlightenment archaeologist C. J. Thomsen saw the "ages of man" advance through millennia of mining, and made the ages their namesake—Golden, Silver, Bronze, and Iron.

Also below the surface rests the bounty of ancient oceans. Not just layer upon layer, but millions upon millions of years of life entombed, of carbon compressed, of sunshine buried, transmogrified, and raised once more to light the earth. Fountains of oil and pyramids of coal pay tribute to the sun god.

In more recent oceans, fishes of kaleidoscopic diversity. Up in the sky, a flock of pigeons a billion birds large that blots out the sky and whose droppings fall like "melting flakes of snow."[1] Away to the west, a shimmering brown blanket of bison, millions strong yet each one a royally robed "monarch of the plains."[2] Under their feet, prairie soils laden with incomparably rich humus to grow food that feeds a nation.

These are how a noir environmental history might begin. Spectacularly. Each beginning a seemingly inexhaustible example of what historian Donald Worster called the wealth of nature.[3] Of what could have been, what *should have been*.

Wealth can't buy you happiness preaches the pitiful cliché. Long before hiring an assassin to dispatch her husband, Maurizio, heir to the fashionable Gucci fortune, Patrizia Reggiani simultaneously embraced and dispatched the cliché, godfather-like. "I would rather weep in a Rolls-Royce than be happy on a bicycle," Reggiani decreed.[4] In prison she stayed true to her values. After a decade Reggiani refused a work release program because she would not humble herself as a waitress outside jail walls.[5]

Perhaps natural wealth inspires more admirable values? And fewer murders? Unlike bullion in a vault, bills in a register, numbers on a statement, or extravagant purses in a boutique, two-hundred-foot towers of white pine surely had the physicality to inspire wonder as well as lust. As would have the equally magnificent and nearly as long Blue Whale, whose infant offspring emerge from the womb into their aquatic cosmos an adorable, rotund three tons. Surely the great iron ranges impressed Hephaestus himself—in antiquity the famed Erzberg ("ore mountain") of Noricum, in modernity the Mesabi of Minnesota, and today, Afghanistan's "astonishing" Hajigak, a deposit of "stunning potential," proclaimed American war general David Petraeus.[6] Surely again, discoveries of immense oil fields in the Gulf of Mexico, the Persian Gulf, the Niger Delta, the Caspian Sea, all brought relief and gratitude to poor communities.

Wouldn't the leaders of such plentiful lands and seas seize the chance to guide their people toward a good life, and to celebrate the land that gave great gifts? Wouldn't these natural wonders inspire care for earthly things beyond oneself—beyond one's human self? Or humility in the face thereof? Most and least of all, wouldn't everyone want these wonders to *last*? And yet, whether ancient, modern, or mythical, coppery bright stories of unrivaled natural wealth often take a turn toward collapse.

Within his five-painting series titled "The Course of Empire," nineteenth-century American artist Thomas Cole offered an allegorical scene of collapse and its aftermath. The penultimate piece is "Destruction," which depicts the sack of an empire's capital and the murder and rape of its citizens. History's real destructions are ugly, vicious, and far more excruciating to study than Cole's exquisitely lit portent for an ascendant America.

Collapses often feature a magnificent forest. Some writers have looked back to the doomed forests of the ancient Mediterranean, and even to prehistoric Easter Island.[7] Oft told in North America are versions of "The Great Cutover," the rapid liquidation of eastern white pine in the upper Great Lakes during the nineteenth century. In the twentieth century it's the Amazon rainforest, a 1.4-billion-acre biological metropolis of living skyscrapers, its four strata of life starting at ground level, the forest floor, then rising through the understory and canopy to the emergent layer, almost twenty stories up. In Cole's depiction of destruction, his painting cannot hold a candle to the smoldering fires and roaring chainsaws that raze *millions* of acres of Amazonia a year: crashes of sylvan giants, cries from families of those threatened then murdered for daring to protect forest homes, a char-gray aftermath. And the future? A frightful apparition of carbon once confined to its forest "sink" but now escaped to heat a too-warm world.

From a forest-green choral to an inky-black beat, the incantation of nature's fabulous wealth builds through the twentieth century:

Spindletop, Bolivar, Chicontepek, Kirkuk,
Gachsaran, East Texas, Agha Jari, Burgan,
Spraberry Trend, Romashkino, Ghawar, Safaniya-Khafji,
Rumaila, Pembina, Block Zero, Bomu,
Hassi Messaoud, Ahwaz, Daqing ("great celebration"),Toot,
Sarir, Mumbai High, Samotlor, Zakum,
Lyontorskoye, Kuparuk, Ekofisk, Prudhoe Bay,
Forties, Fyodorovskoye, West Qurna, Majnoon,
Cantarell, East Baghdad, Hibernia, Tengiz,
Azeri-Chirac-Gunashli, Azadegan, Kashagan, Sugar Loaf.

These were Babel personified. But they shared a common language. Also, they shared a geography, an economy, a timeline. They were Hades-sized pools of oil that could have illuminated the underworld. Erupting from the surface, they were triumphant gushers to usher in an Age of Oil.

Petrolia is a harsh world. Here technology and its owners are ruthless, the land and its people subordinate. Indeed, in the landscapes of Petrolia, people are in the way. Giant gas flares light its oil capitals so fiercely you can see them from space. In the Niger Delta, flares have burned continuously for nearly half a century. They create a poisonous hell on earth for those who live among the fumes and eerie light—a community that is living in fire, "living in bondage," cries a resident.[8] The flares burn off lifetimes of natural gas as if it's trash. The arrogance of such flaunted waste takes your breath away. But what else do these undead flames reveal?

This land was a cornucopia, sustained by Africa's largest wetland and sustaining one of its densest populations. Interlocking curves organized life here: the bends in the Niger River, the upward arc of fronds on raffia and oil palms, and most intricate of all, the mangrove roots. Mangrove roots are spectacularly disorienting. Envision a wooden cascade, or perhaps the overlapping shoots of a fireworks display. Adapted to harsh conditions—salinity, sediment, tides—mangroves are shelter, bed, nursery, and pantry to ocean and freshwater animals. In the delta this included a multitude of fin fish and shellfish—sardines, crabs, clams, oysters, shrimp. Worldwide, seventy-five percent of all the species of fish we buy spend some of their lives in a mangrove swamp. Royal Dutch Shell overwrote this landscape.

In 1956 at Oloibiri, Nigeria, Shell D'Arcy discovered its first commercial oilfield in the region. From Oloibiri, Shell mapped the region's oil wealth by bulldozing and detonating 56,000 kilometers or 35,000 miles of seismic lines (and Shell's competitors that length again); it ran 7,000 kilometers of pipeline alongside villages, through rivers and mangroves, over cassava fields and fish beds.[9] Within decades a petrolian maze squeezed out the space and conditions necessary for life. Multiply the Exxon Valdez or BP Deepwater Horizon disasters by fifty years, then add Shell's collusion with an authoritarian government to foreclose any redress (or prevention), and the horror might be imaginable: 600 million gallons of oil spilled, 50,000 acres of mangrove killed, fish disappeared, water contaminated, air poisoned, pipes exploded, a poet hanged, average population density 1,250 per square mile, average life span forty-three years, average daily wage $1, refugees, repression, protest, uprising, executions, low-grade war. "Oil, blood, and fire."[10]

In the 2007 film *There Will Be Blood*, the face of Daniel Plainview stands in for the faceless corporate ancestors of Shell, BP, Exxon-Mobil, Chevron, and Total, today among the world's richest companies. Set in early 1900s California, proto-oil baron Plainview persuades naïve landowners to sell their land by oozing his own respectability and family values and promising to invest in the community. Time reveals his values to be a mask, his promises an emotional scam. At the climax, Plainview faces his longtime enemy, the repulsive preacher Eli Sunday. Sunday is now broker for the property of Plainview's lone holdout, and desperate for the sale. "I'd like you tell me that you are a false prophet, and that god is a superstition," Plainview demands, before agreeing to the transaction. "Say it like it's your sermon!" In cognitive agony, Sunday wails the words, I am a false prophet, and god is a superstition. Then Plainview reveals all: he got the oil from that property years earlier. "Drainage, Eli, you *boy*," he explains. "Here: If you have a milkshake, and I have a milkshake. And I have a straw. And my straw reaches across the room and starts to drink your milkshake. I. Drink. Your. Milkshake!! I drink it up," he hisses, before crushing Sunday's head with a bowling pin.

There Will Be Blood was based on Upton Sinclair's 1927 novel *Oil!*[11] Sinclair based *Oil!* on the Pan American Petroleum and Transport Company, which itself became part of BP's family tree. Trace the family trees of transnational petroleum giants, and you'll see how they intersect with the family trees of mining, chemical, forestry, agriculture, steel, automobile, railroad, and shipping companies. They were all career immigrants, so you can follow their migration routes across the globe. Novel and film provide fictional bookends to the Age of Oil so far as it's written.

Donald Worster writes about the siren call of nature's abundance in our ecological imagination: for religious dreamers, a natural bequest to bless an earthly Eden; for social dreamers, a natural foundation to undergird an enduring Utopia.[12] Such dreams have always been willful, because they depend on forests too vast to cut down, fish too plentiful to vanish, soils too fertile to go barren, ore veins too rich to play out, oil fields too deep to run dry, people too wise to be stupid. The most seductive dream has been more modest but just as illusory: that a community can successfully trade the greatest of its natural wealth for security, stability, and prosperity.

Depending on the individual history—the time, the place, the plants or animals or earthly materials—some set of interconnected reasons will explain why events unfolded one way and not another. Historians might debate those reasons, add to them, revisit and revise them. At a meta-level, we might show how a society shaped its world via its values and beliefs, and also its systems of knowledge, and how quickly or slowly that society adapted to a world whose changes were partially out of its control. At a micro-level: the way a community organized itself for work and play—class, race, ethnicity, gender—and the patterns of a particular ecosystem or landscape. For nineteenth and twentieth-century environments, the stories will likely involve some combination of relationships to nature, advances in technology, scales of production, movement of workers, habits of consumers, behavior of markets, development of policy and—always—a changing and changeable natural world. These explanations are usually complex. They are often impartial, ingenious, persuasive, provocative, fascinating, maybe even brilliant. But they are not always gut-level *true*. Why does unashamed fiction like *There Will Be Blood* feel *true*? Why does Plainview's scary pantomime of a straw stay so vivid?

As a visual cliché, sure it makes sense in an instant, the straw draining a resource dry. But what really resonates are the deeper levels of knowing the straw reaches. Control is one of those levels—control over place: the struggle for possession, for access, for rights, for a say in how the unique wealth of a place (however you define that wealth) will be used, distributed, safeguarded.

Power is another level of knowing—the peril of unrestrained power exercised, for a time at least, over people and nature alike. Mid-century, sociologist C. Wright Mills coined the phrase "power elite."[13] This was his shorthand for the "corporate rich," the "political directorate," and the "warlords," who, as Mills saw it, had come to hold power at the top rung of American society. These were business, political, and military leaders manning their respective "command posts," but also rotating among them, entirely at ease with a stint in one sector, then another. Were he alive these sixty years later, Mills might reasonably think our shorthand of "the one percent" was a vernacular grandchild of "the power elite." Today, communities blessed or cursed with the greatest of nature's wealth might reasonably believe an international power elite makes decisions that will decide their own future.

Herein lies the paradox: awesome abundance in nature has often assured its decimation. What's more, for everything that has been lost (and sometimes partially recovered), there is still too much to lose. Some say water is the oil of the twenty-first century. Half the world's population endures water shortages. Military analysts project migrations of water refugees. The World Economic Forum says we're on the brink of water bankruptcy. From oil wars to water wars. In this dark watery film, Russia's Aral Sea played out early, America's Ogallala aquifer is in decline, a slow-drip catastrophe not yet averted, and California's Central Valley just emerged from a one-thousand-year megadrought. The Laurentian Great Lakes . . . well, they would be the oilfields of freshwater.[14]

These alone make the Great Lakes worth knowing, its history worth understanding. Not all Great Lakes stories are doom and despairing reactions to doom. At an accidental reef in the St. Clair River, awareness, action, and hope became persistent themes. In the Great Lakes writ large, a century-long legal and diplomatic framework called the Boundary Waters Treaty of 1909 allowed Canada and the United States to anticipate problems, to negotiate shared management, to assume responsibilities for problem solving. In this place that is fragile and formidable, besieged and beloved, we have a few optimistic counterpoints to the paradox of abundance, and for sure the Great Lakes offer a more hopeful storyline than the destruction of the Aral Sea. But the extraction of a place's wealth is something to face head-on, especially when some of that wealth remains. Our predecessors overwhelmed then consumed great landscapes and waterscapes. We have to change the trajectory of those stories.

The Accidental Reef

It's late April on the river bottom, at a reef of coal cinders encrusted with zebra mussels and coated with silt and periphyton (algae, aquatic plants, microorganisms). For a second, freeze this *tableau vivant*. Striped shells skew silent and still on their rocky stage, each off kilter from its umbo, the pivot where both halves of the shell meet. Now, unfreeze the frame, and release the reef to its ravels and intrigues.

It's April at the reef, and then May. Round gobies make their rounds. A common carp roots for crawfish and worms. At the gravelly edge, just below a riffle, a six-inch hornyhead chub works his construction site. He clears gravel, then hollows a nest. His scales are golden, belly beige, fins burnt orange, head like a football helmet, and he sports a seasonal red spot behind the eye to beckon a female. They spawn.

Afterward, the chub carries pebbles and stones to the nest. Stone by stone, he creates a rock garden where his planted progeny can hatch and grow in safety. For added protection, the top of the chub's head sprouts upwards of sixty tubercles, protrusions that look like pointy white warts. The tubercles have the force of brass knuckles if he needs to head-butt an egg predator. More spawning, until the garden becomes a mound. It reaches eight inches high and a splendid two feet across. As the mound builder works then spawns, works then spawns, dozens of smaller fish dart about, spawning on the same nest. They're too tiny to move stones themselves, so they place their eggs under his protection.

Come mid-May, redhorse suckers trickle into the reef until they reach the hundreds. They banish the gobies, then clean half-buried cinders. Spring housekeeping finished, they dig their own nests. But downstream, gobies binge on drifting redhorse eggs or hatched fry.[1]

By the third week in May, lake sturgeon arrive. As female sturgeon broadcast eggs, redhorse suckers follow and feed.[2] When the sturgeon finish their spawn and leave, round gobies return for redhorse and sturgeon eggs, and also for zebra mussels and crayfish.[3] Next in line at the reef: logperch, yellow perch, and then smallmouth bass.[4] Walleye, too. Mid-June, lake sturgeon return for a second spawn.

A wealth of life spins and cycles in the currents and eddies and channels at the reef. *Once, a steamship served a salt mine on the shores of Lake St. Clair, at Algonac* . . . and it left an accidental legacy, a living beacon on the bottom of our inland seas.

Acknowledgments

In these strange days, with family gatherings disrupted, reflecting on a book's debts and fortunes is truly a moment to give thanks.

The Accidental Reef and Other Ecological Odysseys in the Great Lakes is now unimaginable without Glenn Wolff's luminous art. Glenn brings fantastical magic and his own gentle spirit to these stories, and to all of us who are fortunate enough to enter his cosmos.

Likewise, the book is unimaginable without Kathy Johnson and Greg Lashbrook. I'm still awed they took a chance on a stranger who called out of the blue and presumptuously asked to see a river bottom through their eyes. Today you can find Greg and Kathy at Polkadotperch.com.

How to thank Jerry Dennis? Definitely for the beauty of his prose, including a seminal book that revealed our sparkling inland seas to rapt audiences. And especially for sensitive mentorship, warm community building, nuanced feedback. Like many others, I owe Jerry unpayable debts.

I still toast the day Dave Dempsey said yes to a project with me and Dan Macfarlane. Advisor to the IJC, author of a body of Great Lakes writing, and an influential policy maker, Dave still made room for new people. This book is better (and finished) because of his involvement.

My brilliant sometimes coauthor/coeditor Daniel Macfarlane knows well how his energetic support and delight in collaboration make scholarly pursuits cheerier and more meaningful. Thank you, Dan.

At the MSU Press, editors-in-chief Julie Loehr and Catherine Cocks and their talented team brought the book to life during a difficult year. Admiration and gratefulness for Amanda Frost, Erin Kirk, Elise Jajuga, and Kristine Blakeslee. I also benefited from the perceptive insights of two anonymous reviewers.

So much gratitude to colleagues who shared their expertise, and their fascination with the natural world. I could not have depicted many phenomena without their counsel and interventions: Steve Bertman and Devin Bloom (who both consented to appear in the book); Matthew Clysdale, Sharon Gill, Johnson Haas, Steve Hamilton, Steve Kohler, Carla Koretsky, Phyllis Pelletier, Lew Pyenson, Sarah Reding, Wil Reding, Tiffany Schriever, Alison Swan, Hans Van Sumeren, and Maarten Vonhof.

Also for his generosity of expertise: A tribute to David Leander of the Great Lakes Diveshop, who in honor of his beloved late wife Pam shared his deep knowledge of the precious hundred-years-old message in a bottle he found on a dive near Harsen's Island. Dave saved me from an embarrassing error, which is a reminder to say that any remaining mistakes are entirely my own.

Deep appreciation to the Bear River Writers Conference—to Keith Taylor, Tim Tebeau, and all the exuberant faculty who made writing in Hemingway country a joyful act of art-making. And to fellow authors and friends for their abiding *long-term* encouragement, astute advice, and reading or listening to pieces even before the book took shape: Brian Donahue, Sara Gregg, Nancy Langston, Jay Taylor, and Marsha Weisiger; Joe Brandao, Buddy Gray, Ed Martini, and Eli Rubin; Mary Elizabeth Braun and Jamie Lewis; and Bear River pal Jacqueline Courteau.

Fortuitous collaborations led a few chapters down new paths. Thank you, Mike Delp, Robin Lee Berry, Constanza Hazelwood, Horizon Books, Kim Stevens, and WMU Freshwater students for amazing Traverse City forums on water; also Colin Duncan, Jim Feldman, Noah Hall, Alan MacEachern, Andrew Marceille, Graeme Wynn, and many others for our binational project, *Border Flows*; Kathy Brosnan, Will Barnett, and Ann Keating for somehow corralling us all in *City of Lake and Prairie*. Joseph Cialdella and Jennifer Rupp at the Michigan Humanities Council for community-based "Third Coast Conversations: Dialogues about Water in Michigan"; and Claire Campbell for initiating *Pure Michigan*, our environmental history issue of the *Michigan Historical Review*.

I am indebted to many communities at Western Michigan University, beginning with our visionary land stewards who inspired, motivated, and taught in ways that always influenced me: arborists Nick Gooch and Lou Mitchell, landscape managers Steve Root and Darryl Junkins, former natural areas manager Steve Keto, stormwater manager Keith Pung, and ecologist Todd Barkman; along with generations of undaunted students in Biology Club, Students for a Sustainable Earth, and Campus Beet.

Also at WMU, many thanks to Cybelle Shattuck for a formative introduction to the Michigan Environmental Justice Coalition Summit in Flint; James Cousins, the only person I can commune with on Constantine Samuel Rafinesque; Jason Glatz of University Libraries for his cartographic talents; Dorilee Crown and Amanda Hoger for their administrative flair; Sharon Carlson and Lynn Houghton of the Zhang Legacy Collections Center for their archival polymathism; and Eric Denby, Katelin Johnson, and Jack Levy for tenacious research assistance on historical obscura, like a transient nineteenth-century salt factory, or the earliest model of an electric fish zapper.

And finally, I've been incredibly privileged to work or at least to intersect with all of the amazing faculty in the Institute of the Environment and Sustainability, Department of History, University Center for the Humanities, Lee Honors College, and across the hall on the third floor of Wood, the Departments of Biological Sciences and Geography, Environment, and Tourism.

Kwame Dawes generously allowed us to weave lines of his powerful poem "Lightfall" in an illustration. The chapters "Negotiating Abundance and Scarcity" and "Water, Oil, and Fish" appeared in other volumes (excerpted here or edited and updated slightly): respectively, *Border Flows: A Century of the Canadian-American Water Relationship* (University of Calgary Press, 2016); and *City of Lake and Prairie: Chicago's Environmental History* (University of Pittsburgh Press, 2020). Excerpts from *McElligot's Pool* by Dr. Seuss, TM & copyright © by Dr. Seuss Enterprises LP, 1947, renewed 1974. Used by permission of Random House Children's Books, a division of Penguin Random House LLC. All rights reserved.

Support for research, writing, or related programming came from the Canadian Studies Program of the Canadian government; the Network in Canadian History and Environment; the Michigan Humanities Council; and Western Michigan University's Office of the Vice President for Research, College of Arts and Sciences, Institute of the Environment and Sustainability Gwen Frostic Endowment, and the Department of History Burnham Macmillan Endowment.

Meagan Van Stratt's vibrant spirit always helped calm and clear the mind of obstacles, while Tim and Amy Krone's positivity has been a throughline in our lives.

My mom and dad, Judith and Richard, and my brother, Dan, offered deep wells of familial encouragement; and every spring break, my parents gave me a sanctuary room in their home, to work in silence, away from the busyness of university life.

To Mike and Pam, Nancy and Jeff—in ways small and large, and life-changing, you helped launch these journeys through northlands and lake country. And to Sharon, Maarten, Alex, Laura, Kurt, Maddie, Charlie—may our lakeside campfires at the Stone House or beyond always blaze with kinship.

This book is dedicated to my two loves, Phillip and Jake.

Notes

1. For their "body plan," see Nancy Auer, "Form and Function in Lake Sturgeon," in *The Great Lake Sturgeon*, ed. Nancy Auer and Dave Dempsey (East Lansing: Michigan State University Press, 2013), 11–14. There's also this beautiful passage on sturgeon anatomy from Harkness and Dymond: "These ancestral sturgeons lied in what may be called the middle ages of fish life. Like the knights in the middle ages of human history, most of the fish at that time were armoured. This armour they still wear in the form of large, bony shields. Just as the cumbersome armour of our middle ages has been abandoned in favour of freer movement, so modern fish wear smaller, thinner, overlapping scales which also favour freer movement." W. J. K. Harkness and J. R. Dymond, *The Lake Sturgeon: The History of Its Fishery and Problems of Conservation* (Toronto: Department of Lands and Forests, Fish and Wildlife Branch, 1961), 6.

2. J. C. Boase et al., "Movements and Distribution of Adult Lake Sturgeon from Their Spawning Site in the St. Clair River, Michigan," *Journal of Applied Ichthyology* 27 (Suppl. 2, 2011), 58–65; B. A. Manny and G. W. Kennedy, "Known Lake Sturgeon (*Acipenser fulvescens*) Spawning Habitat in the Channel between Lakes Huron and Erie in the Laurentian Great Lakes," *Journal of Applied Ichthyology* 18 (2002), 486–490; M. V. Thomas and R. C. Haas, "Abundance, Age Structure, and Spatial Distribution of Lake Sturgeon, *Acipenser fulvescens*, in the St. Clair System," *Journal of Applied Ichthyology* 18 (2002), 495–501.

3. R. M. Bruch and F. P. Binkowski, "Spawning Behavior of Lake Sturgeon (*Acipenser fulvescens*)," *Journal of Applied Ichthyology* 18 (2002), 574.

4. Bruch and Binkowski, "Spawning Behavior of Lake Sturgeon (*Acipenser fulvescens*)," 574. See also S. Jerrine Nichols, "USGS St. Clair Waterway Report," in *1999 Activities of the Central Great Lakes Bi-National Lake Sturgeon Group*, ed. Tracy D. Hill and Jerry R. McClain (paper presented at the Great Lakes Fishery Commission, Lake Huron Committee Meeting, Ann Arbor, Michigan, March 20–21, 2000, and Lake Erie Committee Meeting, Niagara-On-The-Lake, Ontario, March 29–30, 2000), 30–36. Nichols (p. 32) notes that sturgeon were not afraid of scuba divers, rubbed against them, and "treated the divers as another large sturgeon."

5. Michigan State University, "Background to Lake Sturgeon Egg Stage," in *Lake Sturgeon and Coupled Great Lakes–Tributary Ecosystems: Long-term Ecological Research; Cheboygan River, MI*, http://www.fw.msu.edu/glsturgeon/sturgeon /life-history/egg-stage. Also helpful: D. S. Murray, M. M. Bain, and C. E. Adams, "Adhesion Mechanisms in European Whitefish *Coregonus lavaretus* Eggs: Is This a

Survival Mechanism for High-Energy Spawning Grounds?" *Journal of Fish Biology* 83 (2013), 1221–1233.

6. See recent age estimates of the Acipenseriformes (sturgeons) in Thomas J. Near et al., "Resolution of Ray-Finned Fish Phylogeny and Timing of Diversification," *PNAS* 109, no. 34 (2012): 13698–13703.

7. Or, picture a mass equal to all the ice in Antarctica.

8. North American coasts have larger anadromous species, including the white sturgeon. But the potamodromous lake sturgeon is the largest fish whose life cycle takes place only in freshwater, not a combination of freshwater and saltwater habitats.

9. Verbatim paragraphs from W. Wade Miller, "Sturgeon: Aquaculture Curriculum Guide; Year Two Species Module," preliminary ed. (Alexandria, VA: National Council for Agricultural Education, 1995), 32.

UNDERWATER RASHOMON

1. "The past is a foreign country; they do things differently there" is the opening line of L. P. Hartley's 1953 novel, *The Go Between* (New York: NYRB Classics, 2002). Geographer David Lowenthal used this for his book title, *The Past is a Foreign Country* (New York: Cambridge University Press, 1988).

2. Genetic analyses of European and North American zebra mussels show populations in Lakes Huron and Erie closest to populations in Poland and the Netherlands (i.e., the Baltic Sea and the Rhine River). Zebra mussel establishment likely took "multiple founding source populations and/or that founding events comprised a large number of individuals." Both possibilities account for the ballast water exchanges of Great Lakes ship traffic in Lake St. Clair. C. A. Stepien, C. D. Taylor, and K. A. Dabrowska, "Genetic Variability and Phylogeographical Patterns of a Nonindigenous Species Invasion: A Comparison of Exotic vs. Native Zebra and Quagga Mussel Populations," *Journal of Evolutionary Biology* 15 (2002), 321, 324. Note, however, that studies like Stepien, Taylor, and Dabrowska's are part of unfolding genetic research comparing North American and European populations. See this (slightly) earlier case for the Black Sea origins of North American zebra mussel populations: Jeffrey L. Ram and Robert F. McMahon, "Introduction: The Biology, Ecology, and Physiology of Zebra Mussels," *American Zoologist* 36 (1996), 241.

3. Anthony Ricciardi and Hugh J. MacIsaac, "Recent Mass Invasion of the North American Great Lakes by Ponto-Caspian Species," *TREE* 15:2 (2000), 62–63; Legislative Assembly of Ontario. Committee Transcripts: Standing Committee on Resources Development, January 29, 1991, Zebra Mussels and Purple Loosestrife, Joe Leach testimony, 1044. For an easy introduction to ballast water, see Paul Bruno, "What Is Ballast Water? Understand Ballast Water Systems, Environmental Effects, and Emerging Technology," http://maritime.about.com/od/Ports/a/What-Is-Ballast-Water.htm.

4. Dan Egan offers dramatic firsthand accounts of their discovery and the growing horror of Great Lakes ecologists. Egan, *The Death and Life of the Great Lakes* (New

York: W. W. Norton, 2017), 108–147. See also "20 Years of Zebra & Quagga Mussel Research at NOAA's Great Lakes Environmental Research Laboratory" (Ann Arbor, MI: NOAA, Great Lakes Environmental Research Laboratory, 2008), poster.

5. Thomas F. Nalepa et al., "Changes in the Freshwater Mussel Community of Lake St. Clair: From Unionidae to *Dreissena polymorpha* in Eight Years," *Journal of Great Lakes Research* 22, no. 2 (1996): 354–369.

6. Henry A. Vanderploeg et al., "Dispersal and Emerging Ecological Impacts of Ponto-Caspian Species in the Laurentian Great Lakes," *Canadian Journal of Fisheries and Aquatic Sciences* 59 (2002), 1213; Danielle M. Crosier and Daniel P. Molloy, eds., "Foot and Byssal Threads," US Army Corps of Engineers, http://el.erdc.usace.army.mil/zebra/zmis/zmishelp4/foot_and_byssal_threads.htm; Ram and McMahon, 240.

7. "20 Years of Zebra & Quagga Mussel Research"; Vanderploeg et al., 1221. Ram and McMahon, "Introduction," 240, describe zebra mussel impact on unionids in highly colonized areas as "complete extirpation."

8. For the evolutionary history of *Dreissena polymorpha* in a volatile Ponto-Caspian context, I rely on David F. Reid and Marina I. Orlova, "Geological and Evolutionary Underpinnings for the Success of Ponto-Caspian Species Invasions in the Baltic Sea and North American Great Lakes," *Canadian Journal of Fisheries and Aquatic Sciences* 59 (2002), 1144–1158; and C. A. Stepien et al., "Evolutionary, Biogeographic, and Population Genetic Relationships of Dreissenid Mussels, with Revision of Component Taxa," in *Quagga and Zebra Mussels: Biology, Impacts, and Control*, ed. Thomas F. Nalepa and Don W. Schloesser, 2nd ed. (Boca Raton, FL: CRC Press, 2012), 406.

9. "Titan Tethys was once married to Oceanus, whose translucent waters scarf the broad earth. Their child Pleione couples with sky-lift Atlas—so the story is—and bears the Pleiades." Ovid quoted by Aaron J. Atsma, "The Theoi Project," in Anthony Boyle and Roger D. Woodard, *Fasti* (New York: Penguin Classics, 2004).

10. Andrei Chepalyga, "The Late Glacial Great Flood in the Ponto-Caspian Basin," in *The Black Sea Flood Question: Changes in Coastline, Climate, and Human Settlement*, ed. V. Yanko-Hombach et al. (Dordrecht, Netherlands: Springer, 2007), 119–148.

11. Acknowledgment and huge thanks to my colleague and friend, evolutionary ecologist Sharon Gill, for our long, fascinating conversations on systematics.

12. The phylogeny is from Rüdiger Bieler et al., "Investigating the Bivalve Tree of Life: An Exemplar-Based Approach Combining Molecular and Novel Morphological Characters," *Invertebrate Systematics* 28 (2014), 32–115. Note that the different lengths of branches denote time.

13. Dates of basal splits since Dreissenidae from Gregory W. Gelembiuk, Gemma May, and Carol Eunmi Lee, "Phylogeography and Systematics of Zebra Mussels and Related Species," *Molecular Ecology* 15 (2006): 1033–1050. *Dreissena* "branch tips" from Stepien et al., "Evolutionary, Biogeographic, and Population Genetic Relationships of Dreissenid Mussels, with Revision of Component Taxa," 404; and Gonzalo Giribet, "Bivalvia," in *Phylogeny and Evolution of the Mollusca*, ed. Winston F. Ponder and David R. Lindberg (Berkeley: University of California Press, 2008), 105–142. For species names, I follow the most recent convention of C. A. Stepien et

al. (2012), which differs in some details from earlier literature. Note that *Dreissena rostriformis bugensis* is the quagga mussel.

14. Reid and Orlova, "Geological and Evolutionary Underpinnings for the Success of Ponto-Caspian Species Invasions in the Baltic Sea and North American Great Lakes," 1151; Carol Eunmi Lee and Gregory William Gelembiuk, "Evolutionary origins of invasive populations," *Evolutionary Applications* (2008): 430, 432; C. A. Stepien et al., "Evolutionary, Biogeographic, and Population Genetic Relationships of Dreissenid Mussels, with Revision of Component Taxa," 404; E. M. Harper, J. D. Taylor, and J. A. Crame, eds., *The Evolutionary Biology of the Bivalvia* (London: Geological Society of London, 2001); Thomas W. Therriault et al., "Molecular Resolution of the Family Dreissenidae (Mollusca: Bivalvia) with Emphasis on Ponto-Caspian Species, Including First Report of *Mytilopsis leucophaeata* in the Black Sea Basin," *Molecular Phylogenetics and Evolution* 30 (2004), 481, 485; Gonzalo Giribet and Ward Wheeler, "On Bivalve Phylogeny: A High-Level Analysis of the Bivalvia (Mollusca) Based on Combined Morphology and DNA Sequence Data," *Invertebrate Biology* 121, no. 4 (2002): 289; Z. Fang, "A New Scenario for the Early Evolution of the Bivalvia," in Abstracts and Posters of the International Congress on Bivalvia at the Universitat Autònoma de Barcelona, Spain, 22–27 July 2006, *Organisms Diversity & Evolution* 6, Electronic Supplement 16, part 1 (2006): 31; Mikhail O. Son, "Native Range of the Zebra Mussel and Quagga Mussel and New Data on Their Invasions within the Ponto-Caspian Region," *Aquatic Invasions* 2, no. 3 (2007), 180.

15. Suzanne M. Pyer, Alice J. McCarthy, and Carol Eunmi Lee, "Zebra Mussels Anchor Byssal Threads Faster and Tighter than Quagga Mussels in Flow," *Journal of Experimental Biology* 212 (2009): 2033–2034; Shanna L. Brazee and Emily Carrington, "Interspecific Comparison of the Mechanical Properties of Mussel Byssus," *Biological Bulletin* 211 (December 2006): 263–274: zebra mussel byssal "threads were the strongest, stiffest, least resistant, and fastest to recover" (263); Elizabeth M. Harper, "The Role of Predation in the Evolution of Cementation in Bivalves," *Palaeontology* 34, no. 2 (1991): 455–460.

16. John Phillips, *The General History of Inland Navigation: Containing a Complete Account of All the Canals of the United Kingdom with Their Variations and Extensions according to the Amendments Acts of Parliament to June 1803 and a Brief History of the Canals of Foreign Countries* (London, 1803).

17. Phillips, *The General History of Inland Navigation*, n.p.

18. Phillips, *The General History of Inland Navigation*, iv.

19. Phillips, *The General History of Inland Navigation*.

20. Phillips, *The General History of Inland Navigation*, xii.

21. Phillips, *The General History of Inland Navigation*, 36.

22. Phillips, *The General History of Inland Navigation*, 32.

23. Phillips, *The General History of Inland Navigation*, 25.

24. Halina Lerski, *The Historical Dictionary of Poland, 966–1945* (Santa Barbara, CA: ABC-CLIO, 1996), 399.

25. Robert E. Jones, *Provincial Development in Russia: Catherine II and Jakob Sievers* (New Brunswick, NJ: Rutgers University Press, 1984).

26. Phillips, *The General History of Inland Navigation*, 31.

27. Expansion of zebra mussel range along European canals and waterways during the eighteenth to twentieth centuries from: Anna Stanczykowska, K. Lewandowski, and M. Czarnoleski, "Distribution and Densities of *Dreissena polymorpha* in Poland: Past and Present," in *The Zebra Mussel in Europe*, ed. Gerard van der Velde, Sanjeevi Rajagopal, and Abraham bij de Vaate (Leiden: Backhuys Publishers, 2010), 119, citing Nowak 1971; Son, "Native Range of the Zebra Mussel and Quagga Mussel and New Data on Their Invasions within the Ponto-Caspian Region," 174–184; Marina I. Orlova, Thomas W. Therriault, Pavel I. Antonov, and Gregory Kh. Shcherbina, "Invasion Ecology of Quagga Mussels (*Dreissena rostriformis bugenis*): A Review of Evolutionary and Phylogenetic Impacts," *Aquatic Ecology* 39 (2005): 405, 408, 412; C. A. Stepien, C. D. Taylor, K. A. Dabrowska, "Genetic Variability and Phylogeographical Patterns of a Nonindigenous Species Invasion: A Comparison of Exotic vs. Native Zebra and Quagga Mussel Populations," *Journal of Evolutionary Biology* 15 (2002), 323, figure 4; Yulia A. Kim, L. V. Kwan, G. Demesinova, "Investigation on the Distribution and Biomass of *Mnemiopsis* in the Kazakhstan Sector of the Caspian Sea in 2002," in *Aquatic Invasions in the Black, Caspian, and Mediterranean Seas*, ed. Henri J. Dumont, Tamara A. Shiganova, Ulrich Niermann (Springer Science and Business Media, 2004), 212 on European Canal history and 213–220 on invasive species range extensions; Erkki Leppäkoski, "Living in a Sea of Exotics: The Baltic Case," *Aquatic Invasions in the Black, Caspian, and Mediterranean Seas*, ed. Henri J. Dumont, Tamara A. Shiganova, Ulrich Niermann (Springer Science and Business Media, 2004), 245, table 2 for the Curonian Lagoon sighting.

28. Alfred W. Crosby, *The Columbian Exchange: Biological and Cultural Consequences of 1492* (Westport, CT: Greenwood, 1972; Santa Barbara, CA: Praeger, 2003). Citations refer to the Praeger edition.

29. Crosby followed *The Columbian Exchange* with *Ecological Imperialism: The Biological Expansion to Europe, 900–1900* (1986; repr., New York: Cambridge University Press, 2004). In *Ecological Imperialism*, Crosby argued that Europeans conquered and replaced indigenous peoples in North America and other temperate zones because of biology more than military might—because indigenous species and peoples were evolutionarily unready to resist coevolved co-invaders from Europe. Thirty years later, scientists and historians have more complex understandings of species invasions. Yet "ecological imperialism" remains a core (interdisciplinary) model, a foundation to build on or rebut. For one example of rebuttal, see Jonathan M. Jeschke and David L. Strayer, "Invasion Success of Vertebrates in Europe and North America," *PNAS* 102, no. 20 (2005), 7198–7202. Jeschke and Strayer reject Crosby's thesis as "the imperialism dogma."

30. Karl G. Heider, "The Roshomon Effect: When Ethnographers Disagree," *American Anthropologist* 90, no. 1 (1998), 73–81; Stephen Prince, "The Rashomon Effect," (Criterion Collection 2012), Blu-ray Special Edition.

31. Kurosawa blended two Ryunosuke Akatagawa stories, "Rashomon" and "In a Grove." See Ryunosuke Akatagawa, *Rashomon and Other Stories* (New York: W. W. Norton, 1999); also Donald Richie, "Introduction," in *Rashomon*, dir. Akira Kurosawa, ed. Donald Richie (New Brunswick, NJ: Rutgers University Press, 1987), 13. American

audiences are more likely to know Kurosawa's *Seven Samurai,* but environmental film buffs should see Kurosawa's *Dersu Uzala* (1975), a depressing film version of Russian scientist and ethnographer V. K. Arseniev's *Dersu the Trapper* (New York: E. P. Dutton, 1941). Originally published in 1923 in Russian.

32. Scholars also call the gate Rajo-mon. Yuko Tagaya, "Kyoto in Myth and Literature," in *Islands and Cities in Medieval Myth, Literature, and History: Papers Delivered at the International Medieval Congress, University of Leeds, in 2005, 2006, and 2007,* ed. Andrea Grafetstätter, Sieglinde Hartmann, James Michael Ogier (Oxford: Peter Lang, 2011), 121; William H. McCullough, "The Capital and its Society," in *The Cambridge History of Japan: Volume 2. Heian Japan,* ed. Donald H. Shively and William H. McCullough (New York: Cambridge University Press, 1999).

33. Henry A. Vanderploeg et al., "Dispersal and Emerging Ecological Impacts of Ponto-Caspian Species in the Laurentian Great Lakes," *Canadian Journal of Fisheries and Aquatic Sciences* 59 (2002), 1209–1228.

34. Vanderploeg et al., "Dispersal and Emerging Ecological Impacts of Ponto-Caspian Species in the Laurentian Great Lakes," 1210.

35. Vanderploeg et al., "Dispersal and Emerging Ecological Impacts of Ponto-Caspian Species in the Laurentian Great Lakes," 1222, say that, "we can only speculate that the Lake St. Clair region environment closely matches donor regions the Black and Caspian Seas."

FEAST AND FAMINE

1. My model here is the intricate mesh web of the *Dictynidae* spider family. Their web patterns come closest to how I imagine a web of life—a patchwork of interconnected orbs, each with its own hub.

2. For urgency, read Dan Egan's much-needed alarm, *The Death and Life of the Great Lakes* (New York: W. W. Norton, 2017), 108–147, which itself emerged from his influential Pulitzer-Prize-nominated, four-part special report on the Great Lakes for the *Milwaukee Journal Sentinel,* "A Watershed Moment: Great Lakes at a Crossroads," July 26, 2014. I reviewed Egan's book for the journal *Ecology,* and titled the review to best convey Egan's message and intense public outreach: "Burning Waters: With the Great Lakes, This Is No Time for Patience."

3. Kornis et al. noted that round gobies did not eat enough zebra mussels to reduce their numbers. M. S. Kornis, N. Mercado-Silva, and N. J. Vander Zanden, "Twenty Years of Invasion: A Review of Round Goby *Neogobius melanostomus* Biology, Spread and Ecological Implications," *Journal of Fish Biology* 80, no. 2 (2012), 235–285; Henry A. Vanderploeg et al., "Dispersal and Emerging Ecological Impacts of Ponto-Caspian Species in the Laurentian Great Lakes," *Canadian Journal of Fisheries and Aquatic Sciences* 59 (2002), 1213, 1221; D. J. Jude, Robert R. Reider, and Gerald R. Smith, "Establishment of Gobiidae in the Great Lakes Basin," *Canadian Journal of Fisheries and Aquatic Sciences* 40 (1992), 416–421.

4. *Diporeia* populations crashed in all the lakes except Superior. Andrew M. Muir et al., "Reproductive Life-History Strategies in Lake Whitefish (*Coregonus clupeaformis*) from the Laurentian Great Lakes," *Canadian Journal of Fisheries*

and Aquatic Sciences 71, no. 8 (2014): 1256–1269; Seth J. Herbst, J. Ellen Marsden, and Brian F. Lantry, "Lake Whitefish Diet, Condition, and Energy Density in Lake Champlain and the Lower Four Great Lakes Following Dreissenid Invasions," *Transactions of the American Fisheries Society* 142, no. 2 (2013): 388–398; Richard P. Barbiero et al., "Trends in *Diporeia* Populations across the Laurentian Great Lakes," *Journal of Great Lakes Research* 37, no. 1 (2011): 9–17; Thomas F. Nalepa et al., "Lake Whitefish and *Diporeia* spp. in the Great Lakes: An Overview," in *Proceedings of a Workshop on the Dynamics of Lake Whitefish (Coregonus clupeaformis) and the Amphipod Diporeia spp. in the Great Lakes,* ed. L. C. Mohr and T. F. Nalepa (Ann Arbor: GLFC, 2005), 66. Again, fish did not consume enough zebra mussels to reduce mussel populations.

5. Kornis, Mercado-Silva, and Vander Zanden's review shows that the round goby stimulates heavy metal recycling regardless of existing contamination levels in sediment, but recycling of PCBs and other organic pollutants depends on those ambient levels. Kornis, Mercado-Silva, and Vander Zanden, "Twenty Years of Invasion," 267; Jeffrey L. Ram and Robert F. McMahon, "Introduction: The Biology, Ecology, and Physiology of Zebra Mussels," *American Zoologist* 36 (1996), 240, citing MacIsaac et al. 1996; Vanderploeg et al., "Dispersal and Emerging Ecological Impacts of Ponto-Caspian Species in the Laurentian Great Lakes," 1216.

6. Vanderploeg et al., "Dispersal and Emerging Ecological Impacts of Ponto-Caspian Species in the Laurentian Great Lakes," 1209–1224.

7. *Dreissenids* magnify Lake Erie's already high nutrient load from agricultural runoff. H. A. Vanderploeg et al., "Role of Selective Grazing by *Dreissenid* Mussels in Promoting Toxic *Microcystis* Blooms and Other Changes in Phytoplankton Composition in the Great Lakes," in *Quagga and Zebra Mussels: Biology, Impacts, and Control*, 2nd ed., ed. T. F. Nalepa and D. W. Schlosser (Boca Raton, FL: CRC Press, 2013), 509–523.

8. Dan Egan, *The Death and Life of the Great Lakes* (New York: W. W. Norton, 2017), 300–321; and Dan Egan, "A Great Lake Revival," *Milwaukee Journal Sentinel,* December 7, 2014. Also, Todd A. Hayden et al., "Acoustic Telemetry Reveals Large-Scale Migration Patterns of Walleye in Lake Huron," *PLoS ONE* 9, no. 12 (2014), https://doi:10.1371/journal.pone.0114833.

9. I was with my class at Sleeping Bear Dunes National Lakeshore fall 2006, a year when 3,000 dead birds littered the beach. We saw many dead loons. Outbreaks are worst in years of low water levels and warm temperatures. Kornis, Mercardo-Silva, and Vander Zanden, "Twenty Years of Invasion," 250.

10. "The quagga mussel destruction is so profound it is hard to fathom," says Dan Egan. Egan, *The Death and Life of the Great Lakes*, 123. For an overview of invasional meltdown in the Great Lakes—and a chronology of scientists' distress once a meltdown gets underway—see Jeff Alexander, *Pandora's Locks: The Opening of the Great Lakes–St. Lawrence Seaway* (East Lansing: Michigan State University Press, 2009), 153–165. On the bloody red shrimp, see Anthony Ricciardi, S. Avlijas, and J. Marty. "Forecasting the Ecological Impacts of the *Hemimysis anomala* Invasion in North America: Lessons from other Freshwater Mysid Introductions," *Journal of Great Lakes Research* 38, Supplement 2 (2012): 7. David F. Reid and Marina I.

Orlova, "Geological and Evolutionary Underpinnings for the Success of Ponto-Caspian Species Invasions in the Baltic Sea and North American Great Lakes," *Canadian Journal of Fisheries and Aquatic Sciences* 59 (2002):1146, detail three explanatory "invasion model frameworks" for the Great Lakes: (1) the historical model, in which the life histories of nonindigenous and indigenous species affect ecological outcomes; (2) the vulnerability model, in which ecosystem characteristics like diversity or donor–recipient match make nonindigenous species more or less successful in a new place; and (3) the propagule model—the frequency and intensity of species "supplied" to an ecosystem. All three fit the environmental history of Lake St. Clair. Scientific names and taxonomic groups from USGS's Nonindigenous Aquatic Species Database, 2012, http://nas.er.usgs.gov: quagga mussel, *Dreissena rostriformis bugensis* (mollusk-bivalve); scud, *Echinogammarus ischnus* (crustacean-amphipod); spiny and fishhook waterfleas, *Bythotrephes longimanus* and *Cercopagis pengoi* (crustaceans-cladocerans); bloody red shrimp, *Hemimysis anomala* (crustacean-mysid). A terrific searchable database is NOAA's "Glansis: Great Lakes Aquatic Nonindigenous Species Information System," http://www.glerl.noaa.gov /res/Programs/glansis/glansis.html.

11. Reduced rates of nutrient cycling or nutrient depletion is called oligotrophication. Lakes Michigan and Huron are becoming oligotrophic, and more closely resembling Lake Superior. Caroline Mosley and Harvey Bootsma, "Phosphorus Recycling by Profunda Quagga Mussels (*Dreissena rostriformis bugensis)* in Lake Michigan," *Journal of Great Lakes Research* 41, supplement 3 (2015), http://dx.doi.org/10.1016 /j.jglr.2015.07.007; Steven A. Pothoven and Gary L. Fahnenstiel, "Recent Change in Summer Chlorophyll A Dynamics of Southeastern Lake Michigan," *Journal of Great Lakes Research* 39, no. 2 (2013): 287–294; Richard P. Barbiero, Barry M. Lesht, and Glenn J. Warren, "*Convergence of Trophic State and the Lower Food Web in Lakes Huron, Michigan and Superior,*" *Journal of Great Lakes Research* 38, no. 2 (2012): 368–380; YoonKyung Cha et al., "Do Invasive Mussels Restrict Offshore Phosphorus Transport in Lake Huron?" *Environmental Science and Technology* 45, no. 17 (2011): 7226–7231; Julia L. Mida et al., "Long-Term and Recent Changes in Southern Lake Michigan Water Quality with Implications for Present Trophic Status," *Journal of Great Lakes Research* 36, Special Issue Supplement 3 (2010): 42–49; Henry A. Vanderploeg et al., "*Dreissena* and the Disappearance of the Spring Phytoplankton Bloom in Lake Michigan," *Journal of Great Lakes Research* 36, Special Issue Supplement 3 (2010): 50–59; and most recently, Jiying Li et al., "Benthic invaders control the phosphorus cycle in the world's largest freshwater ecosystem," *PNAS* 118, no. 6 (2021), e2008223118. For quagga mussel population trends in Lake Michigan, see Thomas F. Nalepa, David Fanslow, and Gregory Lang, "Transformation of the Offshore Benthic Community in Lake Michigan: Recent Shift from the Native Amphipod *Diporeia* spp. to the Invasive Mussel *Dreissena rostriformis bugensis,*" *Freshwater Biology* 54 (2009): 475; Thomas Nalepa quoted in Jim Bloch, "Invasion of the Great Lakes: Quagga Mussels Least Known, Most Dangerous Invader," *The Voice*, January 17, 2012; Charles P. Madenjian et al., Appendix 1 in "Status and Trends of Prey Fish Populations in Lake Michigan, 2013" (paper presented at the Great Lakes Fishery Commission, Lake Michigan Committee, Meeting, Windsor,

Ontario, March 25, 2014). In addition, a University of Michigan research team is developing a DNA-based method to quantify dreissenid mussel populations. Finally, aquatic ecologist Steve Kohler points out that many ecologists consider Lake Michigan a simple food web getting simpler, analogous to a giant farm pond (personal communication, July 30, 2015).

ON NAMING AND KNOWING

1. Conversational forumese transcribed exactly. "Fishing Report Port Huron off the Wall," *Michigan Sportsman.com*, posts from March 31, 2010, http://www.michigan-sportsman.com/forum/threads/fishing-report-port-huron-off-the-wall.330784/.

2. Philip Keillor and Elizabeth White, *Living on the Coast: Protecting Investments in Shore Property on the Great Lakes* (Detroit, MI: United States Army Corps of Engineers, Detroit District; Madison, WI: University of Wisconsin Sea Grant Institute, 2003).

3. This is 10 percent of the actual number spawning in Lake Erie

4. "Steelie" refers to Steelhead Trout, and "coho" to Coho Salmon. For how unique fishing the St. Clair River is, see Jerry Dennis's description in *The Living Great Lakes* (New York: Thomas Dunne Books, St. Martin's Press, 2003), 139–40. As Dennis explains it, St. Clair River anglers who fish from boats developed river-specific techniques "designed to get around the challenges of deep water and strong current."

5. Michael Veine, "Erie's Spring Hog Walleyes," Water-N-Woods News, April 1, 2014, http://www.woods-n-waternews.com/Articles-In-This-Issue-i-2014-04-01-217942.112113-Eries-spring-Hog-Walleyes.html

6. Culled from a local fishing forum board.

7. But for those readers who need a kindred angler spirit, or the life-time expertise and first-hand sensations of one: I recommend Jerry Dennis's gorgeous, detailed observations on fishing for walleye in the Huron-Erie corridor, and in Lake Erie ("Walleye World"), along with his survey of the larger ecological history enveloping this and other aquatic species. See Jerry Dennis, *The Living Great Lakes* (New York: Thomas Dunne Books, St. Martin's Press, 2003), 139, 157–163, 167.

8. *Merriam-Webster Dictionary: New Edition* (Springfield, MA: Merriam-Webster, 2016).

9. C. T. Onions, ed., *The Oxford Dictionary of English Etymology* (New York: Oxford University Press, 1966), 990; also, T. F. Hoad, ed., *The Concise Oxford Dictionary of English Etymology* (New York: Oxford University Press, 1993), 532.

10. This last again from Onions, *The Oxford Dictionary of English Etymology*, 990; or Hoad, *The Concise Oxford Dictionary of English Etymology*, 532. For permutations see *Merriam-Webster Dictionary*.

11. Or, *wawilezed* or *waugleeghed*.

12. Ichthyologist Tarleton H. Bean compiled the most impressive catalogue of regional names for walleye in Tarelton H. Bean, *Catalogue of the Fishes of New York* (Albany, NY: New York State University, 1903), 494–495. To my list, add the "most unsuitable" name for the fish, according to Bean, Pennsylvania's "Susquehanna salmon." And more from Bean: the jack (Ohio Valley and western North Carolina),

the white or jack salmon (Ohio Valley), *okow* (Cree Indians), and hornfish (British American fur traders).

13. Blue pike is actually more confusing, since there was a now extinct fish in Lake Erie whose classification could have been walleye itself but with distinct coloring, the way the same species of squirrel might have different colors in different places; or, it could have been a separate species from walleye; or, the ultimate but still unproven conclusion that blue pike was a subspecies of walleye. The question of blue pike still holds interest for Great Lakes taxonomy, for instance, Wendylee Stott et al., "Genetic Diversity among Historical Collections of Two Sander Species from Lake Erie: Comparisons of Microsatellite DNA and SNP Data" (unpublished paper presented at the American Fisheries Society 144th Annual Meeting, Quebec City, Quebec, August 21, 2014).

14. Ichthyology is the science of fish, or the study of fish. Bean, *Catalogue of the Fishes of New York*, 496.

15. This membrane of cells is called the *tapetum lucidum,* a name itself that glimmers, translated as "bright tapestry."

16. For a neat ongoing project on the etymology of scientific fish names, see Christopher Scharpf and Kenneth J. Lazara, *The ETYFish Project: Fish Name Etymology Database,* www.etyfish.org. *Sander vitreus* isn't in the database yet.

17. "Original Communications: Memoir on Ichthyology," *American Monthly Magazine and Critical Review* 2, no. 4 (February 1818), 241–248, https://babel.hathitrust.org /cgi/pt?id=nyp.33433081752622&view=2up&seq=6.

18. *American Monthly Magazine and Critical Review* 2, no. 4 (February 1818), 241.

19. *American Monthly Magazine and Critical Review* 2, no. 4 (February 1818), 247. The quote continues: "Found in the Cayuga Lake, of a roundish (teres) figure; the middling magnitude about eighteen inches long, by three and a quarter deep. . . . This character and description were taken from the drawing and notes of Simeon De Witt, Esq. made by him at Ithaca, in October, 1816; was pronounced by that gentleman to be tolerably good eating." Notice then that namers, including Linnaeus himself, were often not the original collectors and illustrators of the organism.

20. Richard Conniff, *The Species Seekers: Heroes, Fools, and the Mad Pursuit of Life on Earth* (New York: W. W. Norton, 2011).

21. Philosopher and historian of science and medicine Jerry Stannard called the Latin name "a kind of birth certificate to accompany its entrance into the literary botanical world and into which it would, if not illegitimate, find its accepted place." Jerry Stannard, "Linnaeus, Nomenclator Historicusque Neoclassicus," in *Contemporary Perspectives on Linnaeus,* ed. John Weinstock (Lanham, MD: University Press of America, 1985), 24.

22. I am indebted to a number of Linnaean scholars, translators, and other historians of science: Carl von Linne, *Linnaeus' Philosophia Botanica,* trans. Stephen Freer (New York: Oxford University Press, 2003). Paul Alan Cox, "Introduction," in *Linnaeus' Philosophia Botanica,* xv–xxv. Multiple contributors in Andrew Polaszek, ed., *Systema Naturae 250: The Linnean Ark* (New York: CRC Press, 2010): Edward O. Wilson, "The Major Historical Trends of Biodiversity Studies," 1–3; James

Dobreff, "Daniel Rolander: The Invisible Naturalist," 11–28; Gordon McGregor Reid, "Taxonomy and the Survival of Threatened Animal Species," 29–52; Quentin D. Wheeler, "Engineering a Linnaean Ark of Knowledge for a Deluge of Species," 53–61; Zhi-Quiang Zhang, "Reviving Descriptive Taxonomy after 250 Years: Promising Signs from a Miga-Journal in Taxonomy," 95–115; David J. Patterson, "Future Taxonomy," 117–126; James Hanken, "The Encyclopedia of Life: A New Digital Resource for Taxonomy," 127–135; Andrew Polaszek and Ellinor Michel, "Linnaeus-Sherborn-ZooBank," 163–172; Richard L. Pule and Ellinor Michel, "ZooBank: Reviewing the First Year and Preparing for the Next 250," 173–184; Benoît Dayrat, "Celebrating 250 Dynamic Years of Nomenclatural Debates,"185–239. Also see this interview with James Hanken, "Organizing the World in the Age of DNA," *Wired*, May 23, 2007. Multiple contributors in John Weinstock, ed., *Contemporary Perspectives on Linnaeus* (New York: University Press of America, 1985): Vernon Heywood, "Linnaeus: The Conflict Between Science and Scholasticism," 1–16; Jerry Stannard, "Linnaeus, Nomenclator Historicusque Neoclassicus," 17–36; David Hull, "Linné as an Aristotelian," 37–54; Mary Winsor, "The Impact of Darwinism upon the Linnaean Enterprise, with Special Reference to the Work of T. H. Huxley," 55–84; Lars Gustafsson, "Linnaeus and his *Nemesis divina* from a Philosophical Perspective," 117–134; Walter Wetzels, "Goethe and Linné," 135–151; Gunnar Broberg, "Linnaeus's Classification of Man," 153–182; Tore Frängsmyr, "Linnaeus in His Swedish Context," 183–193. Multiple contributors in Tore Frängsmyr, ed., *Linnaeus: The Man and His Work* (Berkeley, University of California Press, 1983): Sten Lindroth, "The Two Faces of Linnaeus," 1–62; Gunnar Eriksson, "Linnaeus the Botanist," 63–109. Heinz Goerke, *Linnaeus, trans.* Denver Lindley (New York: Scribner, 1973). Wilfrid Blunt, with the assistance of William T. Stearn, *The Compleat Naturalist: A Life of Linnaeus* (New York: Viking, 1971). James L. Larson, *Reason and Experience: The Representation of Natural Order in the Work of Carl von Linné* (Berkeley: University of California Press, 1971). Also see the excellent Richard Conniff, *The Species Seekers: Heroes, Fools, and the Mad Pursuit of Life on Earth* (New York: W. W. Norton, 2011).

23. Or *nomenclator* (name giver). David Quammen, "Linnaeus: A Passion for Order," in *Systema Naturae 250: The Linnaean Ark, ed.* Andrew Polaszek (New York: CRC Press, 2010), 5–9.

24. Carolus Linnaeus, *Systema Naturae: 1735 Facsimile of the First Edition; With an Introduction and a First English Translation of the "Observationes,"* ed. and trans. M. S. J. Engel-Ledeboer and H. Engel (Nieuwkoop, Netherlands: B. de Graaf, 1964). Wilfrid Blunt, *Linnaeus: The Compleat Naturalist (Princeton, NJ: Princeton University Press, 2003).*

25. Stannard, "Linnaeus, Nomenclator Historicusque Neoclassicus," 20. Heywood, "Linnaeus," 1–5.

26. A reform Linnaeus first fully applied in *Species Plantarum* (1753). Stannard, "Linnaeus, Nomenclator Historicusque Neoclassicus," 30. The application of binomial nomenclature to animals starts in the 10th edition of *Systema Naturae* (1758). See Andrew Polaszek, "Preface," in *Systema Naturae 250: The Linnaean Ark, ed.* Andrew Polaszek (New York: CRC Press, 2010), vii.

27. Heinz Goerke, *Linnaeus* (New York: Scribner, 1973), 149–155; Blunt, *The Compleat Naturalist: A Life of Linnaeus*, 183–192.

28. Rafinesque holds a special place in the history of taxonomy and early naturalists. He became notorious for creating taxonomic chaos (or taxonomic pollution) by manically naming over 6,000 species and genuses during his lifetime. His critics declared that decades of taxonomic do-overs were necessary to correct his sloppy, wrong classifications. Rafinesque is undergoing rehabilitation today as a complicated scientific visionary rather than a taxonomic madman, and also as a historical figure whose posthumous error-ridden biographical sketch was repeated, with errors intact, up to the present. For a hilarious overview of the seething, see Ansel Payne, "Why Do Taxonomists Write the Meanest Obituaries," *Nautilis* 35 (April 7, 2016). The definitive Rafinesque biography (and determined correction of the historical record) is Charles Boewe, *Life of Rafinesque: A Man of Uncommon Zeal* (Philadelphia: American Philosophical Society, 2011). See also, *Profiles of Rafinesque*, ed. Charles Boewe (Knoxville: University of Tennessee Press, 2003); Conniff, *The Species Seekers*, 111–141. For revealing sketches of Rafinesque's prolific publication while on the faculty of Transylvania University in Kentucky, see James Cousins, *Horace Holley: Transylvania University and the Making of Liberal Education in the Early American Republic* (Lexington: University Press of Kentucky, 2016), 138, 163, 182. A final factoid: Rafinesque named *Acipenser fulvescens*, lake sturgeon.

29. Regarding Cuvier's name, *Lucioperca americana*, or American Lucioperca: One Latin derivative of "lucio" means pike, so Lucioperca was pike-perch. But wait, another Latin derivative of "lucio" means bright, luminous, shiny.

30. Louis Agassiz, *Lake Superior: Its Physical Character, Vegetation, and Animals, Compared with Those of Other and Similar Regions* (Boston, 1850); Louis Agassiz, *Contributions to the Natural History of the United States of America* (Boston, 1857–1862).

31. For a fascinating explanation of how Agassiz made himself the central character in his own scientific publications by dividing authorship with people like Cabot, see Christoph Irmscher, *Louis Agassiz: Creator of American Science* (New York: Houghton Mifflin Harcourt, 2013), 95–96.

32. Christoph Irmscher, "The Ambiguous Agassiz," *Humanities* 34, no. 6 (November/December 2013): n.p.

33. The Agassiz party included nine students (the sciences and law principally represented), two Bostonians (including Cabot), three naturalists (addressed as Drs. in the preface), and a geologist from France. Agassiz, *Lake Superior*, iii.

34. The full quote in all its ennui: "In spite of all the glorification on the score of the 'Great Lakes,' it must be confessed that the lower lakes at least are only geographically or economically great. Any one accustomed to the sight of the ocean has to keep in mind the square miles of extent, to preserve his respect for them. Their waves, though dangerous enough to navigators, have not sufficient swing to carve out a rocky shore for themselves, or to tumble any rollers along the beach, and thus the line, where land and water meet, in which, as has been well said, the interest of a sea-view centres, is as tame as the edge of a duck-pond. Much of this

character is doubtless owing to the flat prairie country by which they are mostly surrounded." Agassiz, *Lake Superior*, 20. By contrast, Cabot's felt wonder in the wilderness of Lake Superior, as synopsized by John Knott, *Imagining the Forest: Narratives of Michigan and the Upper Midwest* (Ann Arbor: University of Michigan Press, 2012), 21–23.

35. Agassiz, *Lake Superior*, 21. For a fascinating article on the geological formation of the St. Clair River delta, see Richard L. Thomas et al., "Formation of the St. Clair River Delta in the Laurentian Great Lakes System," *Journal of Great Lakes Research* 32, no. 4 (2006): 738–748.

36. Anne-Marie Oomen, ed., *Elemental: A Collection of Michigan Creative Nonfiction* (Detroit: Wayne State University Press, 2018).

37. Constance Fenimore Woolson, *Castle Nowhere: Lake Country Sketches* (1875; repr., Ann Arbor: University of Michigan Press, 2004), 210.

38. Woolson, *Castle Nowhere*, 226.

39. Woolson, *Castle Nowhere*, 210.

40. *Hearing before the Resources House of Representatives: H.R. 2822, to Reaffirm and Clarify the Federal Relationship of the Swan Creek Black River Confederated Ojibwa Tribes as a Distinct Federally Recognized Indian Tribe, and for Other Purposes*, 105th Cong. (1998) (prepared statement of Deborah Davis Jackson, PhD), (Washington, DC: US Government Printing Office, 1998), 48. Highly recommended: The entire hearing is a contentious, fascinating mosaic of competing histories and contemporary political and economic conflict among American Indian groups and between them and federal or state agencies and governments.

41. Keith R. Widder, *Battle for the Soul: Métis Children Encounter Evangelical Protestants at Mackinaw Mission, 1823–1837* (East Lansing: Michigan State University Press, 1999).

42. Agassiz, *Lake Superior*, 23.

43. Lynn Armitage, "Mackinac Island Finally Telling Native Side of History: Agatha Biddle's Home Will Be Restored to Tell Native History of Mackinac Island," *Indian Country Today* (Verona, NY), March 30, 2017; Michael A. McDonnel, *Masters of Empire: Great Lakes Indians and the Making of America* (New York: Hill and Wang, 2015); Keith R. Widder, *Battle for the Soul: Métis Children Encounter Evangelical Protestants at Mackinaw Mission, 1823–1837* (East Lansing: Michigan State University Press, 1999), 51–53.

44. Agassiz, *Lake Superior*, 22–24.

45. Agassiz, *Lake Superior*, 24.

46. Agassiz, *Lake Superior*, 24.

47. Georges Cuvier and Achille Valenciennes, *Histoire Naturelle Des Poissons*, vol. 2 (Paris, 1828), 122. For the specific illustration, see General Research Division, New York Public Library, "Sandre d'Amérique (P. Lucioperca Americana. N.)," New York Public Library Digital Collections. For an annotated translation of Cuvier's opening survey of ichthyology in the first of his twenty-two-volume series, see Georges Cuvier and Theodore W. Pietsch, *Historical Portrait of the Progress of Ichthyology, from Its Origins to Our Own Time (Foundations of Natural History)* (Baltimore, MD: Johns Hopkins University Press, 1995). Here's Cuvier's 1828 description of Le

Sandre D'Amérique (*Luciperca americana, nob.*): "*Les eaux des États-Unis possèdent un sander qui réunit aussi plusieurs des caractères de la perche.* [The waters of the United States possess a sander, which also combines several of the characteristics of the perch.] *Un peu plus alongé encore que le sander ordinaire, il est partout finement marbré ou réticulé de noirâtre sur un fond jaunâtre ou verdâtre: il a une pointe aiguë à l'opercule, ce qui le différencie beaucoup des sandres d'Europe, et montre en même temps que cette sorte d'armure ne peut fournir que des caracteres très-secondaires. Sa première dorsale est marquée d'une tache noire comme à la perche. Du reste, par les dents et les autres caractères il ressemble au sander, ayant seulement deux rayons de moins è la seconde dorsale.*"

48. Find the *Sander vitreus* record or look for another species in WoRMS, the World Register of Marine Species: http://www.marinespecies.org/aphia.php?p=taxdetails &id=275311.

49. For the larger issue, see R. E. Froese, E. Capuli, and M. C. Rañola, "Challenges to Taxonomic Information Management: How to Deal with Changes in Scientific Names," in *Global Environmental Researches on Biological and Ecological Aspects* (Tsukuba, Japan: Center for Global Environmental Research, 2000), 3–10. Froese (a fish biologist and developer of FishBase), Capuli, and Rañola use as their core example the issue of 53,000 fishes described from Linnaeus to the present, only half of which have valid names. For the change from *Stizostedion* to *Sander,* see the original in C. Richard Robins and American Fisheries Society, Committee on Names of Fishes, *Common and Scientific Names of Fishes from the United States and Canada*, 5th ed., American Fisheries Society Special Publication 20 (Bethesda, MD: American Fisheries Society, 1991); followed by the change in the next edition, Joseph S. Nelson and American Fisheries Society, Committee on Names of Fishes, "Appendix 1: Changes from 1991 Edition and Comments," *Common and Scientific Names of Fishes from the United States, Canada, and Mexico*, 6th edition, American Fisheries Society Special Publication 29 (Bethesda, MD: American Fisheries Society, 2004). Canadian fishing writer and angler Lonnie King also saw the change, and his announcement made its way to walleye fishing forums. See Lonnie King, "Not a Walleye," *Ontario Out of Doors* 36:1 (February 2004), 11; and the subsequent discussion, "Walleye to get new scientific name," at Walleye Message Central, March 2–5, 2004, http://www.walleyecentral.com/forums/archive/index. php/t-31405.html. Toward the end of the thread, Jeff G. asks, "Why the use of Latin for species identification? If they truelly [*sic*] are going to use the species first designation, it likely was not at all a Latin name. Scientists can really get full of themselves! Capitonis Bullatusum!"

50. For example, Paul Levy, "The Big Fish Story Was on the Menu; When Is a Walleye Not a Walleye? When It's a Zander in Disguise," *Star Tribune* (Minneapolis, MN), December 3, 2004; Rob Zaleski, "Fishy Fries on the Menu? Cheaper Zander Often Substituted," *Capital Times* (Madison, WI), August 9, 2007.

51. Marine scientist Nick Higgs gives a concise definition: "Taxonomy is the science of identifying and describing new species, and falls within the biological discipline

of systematics, which is concerned with classifying organisms according to their evolutionary history." Nick Higgs, "Taxonomy in Trouble? An Ocean Science Perspective," *Ocean Challenge* 21, no. 2 (2016): 10–11.

52. S. Lindroth, qtd. in Heywood "Linnaeus: The Conflict Between Science and Scholasticism," 2. Quoted from S. Lindroth, "Linnaeus in His European Context," *Yearbook of the Swedish Linnaean Society*, commemorative volume (1979), 9–17. Heywood also talks briefly about Linnaeus's work with fishes, relying on A. Wheeler, "The Sources of Linnaeus's Knowledge of Fishes," *Yearbook of the Swedish Linnaean Society*, commemorative volume (1979), 156–211. Edward O. Wilson identifies four historical periods for systematics to the present: (1) Aristotle's hierarchical system (the "Aristotelian mode," so antiquity through Middle Ages and the Enlightenment); (2) the "Linnaean enterprise," which "formalized the hierarchical system," established binomial nomenclature, and to find "the entirety of biodiversity" (eighteenth century on); (3) evolutionary theory as the foundation for biodiversity studies (nineteenth century); and (4) completing the Linnaean enterprise by mapping the world's biodiversity (in its infancy today). The fourth is crucial, says Wilson, because "we live, in short, on a little known planet. When dealing with the living world, we are flying mostly blind. When we try to diagnose the health of an ecosystem such as a lake or forest, for example, in order to save and stabilize it, we're in the position of a doctor trying to treat a patient knowing only 10 percent of the organs." See Edward O. Wilson, "The Major Historical Trends of Biodiversity Studies," in (*Systema Naturae 250: The Linnaean Ark, ed.* Andrew Polaszek (New York: CRC Press, 2010), 3.

53. The Encyclopedia of Life (EOL; eol.org) is the vision and "dream brought to reality" of biologist Edward O. Wilson, to "provide global access to knowledge about life on Earth." Wilson, "The Major Historical Trends of Biodiversity Studies," 3. See also, EOL's "taxonomic backbone" is "The Catalogue of Life," a literal online catalogue: Y. Roskov et al., eds., "Species 2000 and ITIS Catalogue of Life, 27th June 2016," digital resource at www.catalogueoflife.org/col. For Linnaeus and mass production, see Heywood, "Linnaeus: The Conflict Between Science and Scholasticism," 6. Heywood cites G. Eriksson, "The Botanical Success of Linnaeus: The Aspect of Organization and Publicity," *Yearbook of the Swedish Linnaean Society*, commemorative volume (1979), 57–66.

54. Quotation from Tracie Watson, "86 Percent of Earth's Species Still Unknown?" *National Geographic News*, August 25, 2011. For the original study, see Camilo Mora et al., "How Many Species Are There on Earth and in the Ocean?" *PLoS Biology* 9, no. 8 (2011), https://doi.org/10.1371/journal.pbio.1001127; and also, M. J. Costello, Robert M. May, and Nigel E. Stork, "Can We Name Earth's Species before They Go Extinct?" *Science* 339, no. 6118 (2013): 413–416. Nick Higgs underscores the same point with one deep-seabed survey in which "a staggering ~90% of species brought up are new to science!" Nick Higgs, "Taxonomy in Trouble? An Ocean Science Perspective," *Ocean Challenge* 21, no. 2 (2016): 10.

55. Watson, "86 Percent of Earth's Species Still Unknown?"

56. Wilson, "The Major Historical Trends of Biodiversity Studies," 1–3.

57. Quentin D. Wheeler, "Engineering a Linnaean Ark of Knowledge for a Deluge of Species," in *Systema Naturae 250: The Linnean Ark, ed.* Andrew Polaszek (New York: CRC Press, 2010), 53–61.

58. Wheeler, "Engineering a Linnaean Ark of Knowledge for a Deluge of Species," 56.

59. Henry Wadsworth Longfellow, *The Building of the Ship, and Other Poems, Michigan Historical Reprint Series* (Ann Arbor: University of Michigan Library, 2006), 3–23.

60. Photo by Gordon Roberton, in Blunt, *The Compleat Naturalist: A Life of Linnaeus*, 222.

61. For a rigorous historical examination of the boundaries between science and art, see David Topper, "Toward an Epistemology of Scientific Illustration," in *Picturing Knowledge: Historical and Philosophical Problems Concerning the Use of Art in Science, ed.* Brian S. Baigrie (Toronto: University of Toronto Press, 1996), 215–249.

62. David Starr Jordan, "Sketch of Charles A. Le Sueur," *Popular Science Monthly* 49 (1895): 549–550; David Starr Jordan, "The History of Ichthyology," *Science* 16, no. 398 (August 15, 1902): 241–58. (Again on Lesueur as artist and naturalist, p. 256: "Le Sueur's engravings of fishes . . . are still among the most satisfactory representations of the species to which they refer.") Also, Josephine M. Elliott and Jane Thompson Johansen, *Charles-Alexandre Lesueur: Premier Naturalist and Artist* (New Harmony, Indiana: J. T. Johansen, 1999). Lesueur's contemporary Rafinesque acknowledged Lesueur as "the first to explore the Ichthyology of the Great American Lakes" and "the first naturalist who visited Lake Erie and Lake Ontario, where he detected a great number of new species," in Constantine Samuel Rafinesque *Ichthyologia Ohiensis or Natural History of the Fishes Inhabiting the River Ohio and its Tributary Streams* (Lexington, KY, 1820), 4, 6. Both Lesueur and Rafinesque were contemporaries of and had traveled with the American naturalist-painter John James Audubon.

63. For a decent chronology of eighteenth and nineteenth century illustrators and scientific artists, with strong attention paid to ichthyology, see David Knight, *Zoological Illustration: An Essay toward a History of Printed Zoological Pictures* (Hamden, CT: Dawson-Archon Books, 1977). For more holistic and penetrating analyses of illustrations as visual, cultural, and scientific "texts," see American Museum of Natural History, *Natural Histories: Extraordinary Rare Book Selections from the American Museum of Natural History Library, ed.* Tom Baione (New York: Sterling Signature, 2012); and Victoria Dickenson, *Drawn from Life: Science and Art in the Portrayal of the New World* (University of Toronto Press, 1998).

64. Jessica Riskin, "Just Use Your Thinking Pump," review of *The Scientific Method: An Evolution of Thinking from Darwin to Dewey*, by Henry M. Cowles, *New York Review of Books* 67, no. 11 (July 2, 2020): 50.

65. Walter Wetzels, "Some Observations on Goethe and Linné," in *Contemporary Perspectives on Linnaeus, ed.* John Weinstock (New York: University Press of America, 1985), 135–151; Blunt, *The Compleat Naturalist: A Life of Linnaeus*, 214.

66. Christoph Irmscher, "The Ambiguous Agassiz," *Humanities* 34, no. 6 (November/December 2013): n.p.

67. Ralph Waldo Emerson, *Nature* (Boston, 1836), 12–13.

68. Kaveh Akbar, "'I'm Not Sorry for Writing about Wonder and Joy.' Aimee Nezhukumatathil," *Divedapper*, June 29, 2015.

69. Devin Bloom, personal communication with the author, August 2, 2017.

70. For an example of such synergy in the classroom, see poet Sandy Feinstein and molecular biologist Bryan Shawn Wang's beautiful "report" on their co-taught class. "Editing," says Feinstein, reflecting on their students practicing gene transfer with recombinant DNA technology, "it's what molecular biologists and poets do" (71). "Field Notes & Marginalia," in *Scientists and Poets Resist, ed.* Sandra L. Faulkner and Andrea England (Boston, Massachusetts: Brill Sense, 2020), 63–71.

71. Poets Keith Taylor and Michael Delp are contemporary role models. Taylor had a long-time summer appointment at the University of Michigan's LSA Biological Station, while Delp has worked closely with Freshwater Science students at Western Michigan University. For a brief sample of Taylor, see "Anne-Marie Oomen & Keith Taylor on the Poetry of Science," *Nature Change: Conversations about Conservation and Climate,* September 22, 2016; Sara Adlerstein-Gonzalez et al., "Mapping the River: A Multimedia Performance about the Cycle of Water and the Relationship of Culture and Water Told Using the Huron River as a Model," performance at the Society for Freshwater Science annual meeting, May 21, 2018, Detroit, Michigan; Keith Taylor, *The Bird-while* (Detroit, MI: Wayne State University Press, 2017). And Delp: "Two Writers and the River: A Conversation with Jerry Dennis & Michael Delp," *Nature Change: Conversations About Conservation and Climate,* May 26, 2016; Michael Delp, *Lying in the River's Dark Bed: The Confluence of the Deadman and the Mad Angler* (Detroit, MI: Wayne State University Press, 2016); and an early tactile beauty I treasure called *The Text of the River* (Farmington, MI: Riverwatch, 1993)—with exquisite artwork by printmaker Ladislav R. Hanka, and bound by legendary bookbinder Jan Sobota.

BAD DIVER

1. Great Lakes Divecenter owner Dave Leander discovered the bottle on a dive, to the envy of his diving friends. He keeps the bottle as a treasure in honor of his beloved late wife Pamela Leander, co-owner of the Divecenter and also an expert diver. Detroiters Selina Pramstaller and Tillie Esper were on White Star Line's famous flagship steamer *Tashmoo*, named for Tashmoo Park. Dave explains that they wrote their message on a deposit slip for beer or pop containers. Ship patrons would redeem their deposits when they returned the containers. Dave Leander, personal communication with the author, November 6, 2020. For a historical recreation of the women's day, see Jim Bloch, "Tashmoo 'message in a bottle' turns 100," *The Voice* (Clinton Township, Michigan), July 2, 2015. As for Tashmoo Park itself, this "was the place to go, the place to be when the waterways were highways," said Harsens Island St. Clair Flats Historical Society president Bernard Licata to the *Detroit Free Press*, "Tashmoo Days on Harsens Island Offer a Trip Back in Time," July, 16, 2015. As a sidenote: the likely namesake for *Tashmoo* is Lake Tashmoo on the island of Martha's Vineyard (off of Cape Cod, in Massachusetts).

2. Even in 1855. "Detroit Archdiocese Says Missing Church Found under Lake St. Clair," Associated Press, December 9, 1995.

1. Today you can find Greg Lashbrook and Kathy Johnson at polkadotperch.com. Unless otherwise indicated, dialogue and quotes come from the author's field visit and interviews with Kathy and Greg, Friday, December 4–Sunday, December 6, 2015. Interview quotations are lightly edited and condensed for clarity.

2. I buy the mask (anyway).

3. "Popping" is when a diver rises too quickly for the body to adjust to the changing air pressure from deeper water to the surface. A quick ascent is dangerous and can even kill a diver.

4. Tom Verdin, "Couple to Take Plunge to Tie Knot," *The Times Herald* (Port Huron, MI), August 2, 1990. In the decades since the Johnson-Lashbrook wedding, underwater ceremonies have become a thing. You can find them easily by entering some kind of wet wedding pun in your Google search.

5. For the unexpected "opportunities" of commercial diving, see Kathy Johnson, "Down in the Dumps: A Day in the Life of a Garbage Dump Diver," *Alert Diver Online*, 2020, http://www.alertdiver.com/garbage-dump-diving.

6. Kathy Johnson and Greg Lashbrook, *Diving and Snorkeling Guide to the Great Lakes: Lake Superior, Michigan, Huron, Erie and Ontario* (Newbury, UK: Pisces), 1991.

7. For people who don't know what "commercial" divers do, here's the Commercial Diving and Marine Services company portfolio: "salvage operations, preconstruction surveys, endangered species investigations, sediment sampling, pipeline inspections, zebra mussel removal, intake cleaning and maintenance of industrial and municipal facilities throughout the Great Lakes."

8. Peter J. Badra, "Status of Native Freshwater Mussels, Including the Northern Riffleshell (*Epioblasma torulosa rangiana*) and Rayed Bean (*Villosa fabalis*), in Detroit River, Michigan" (Michigan State University Extension Report Number 2009-5, Lansing, Michigan, May 5, 2009), 1.

9. The site was near the Manoogian Mansion, the mayor's residence and a Detroit landmark.

10. Gregory A. D., *Manistee Nmé: A Lake Sturgeon Success Story*, 2011.

11. Jimmie Mitchell "N'me," in *The Great Lake Sturgeon*, ed. Nancy Auer and Dave Dempsey (East Lansing: Michigan State University Press, 2013), 24.

12. LRBOI Natural Resources Department and the Nmé Cultural Context Task Group, "Nmé (Lake Sturgeon) Stewardship Plan for the Big Manistee River and 1836 Reservation" (Manistee, MI: Little River Band of Ottawa Indians), n.d.; Marty Holtgren, "Bringing Us Back to the River," in *The Great Lake Sturgeon*, Nancy Auer and Dave Dempsey (East Lansing: Michigan State University Press, 2013), 138–141.

13. Greg was repulsed enough about the domination and force of this procedure that he described it as "raping them [the sturgeon] for the eggs."

14. Holtgren, "Bringing Us Back to the River," 141.

15. Louise Erdrich, *Books and Islands in Ojibwe Country: Traveling through the Land of My Ancestors* (New York: Harper Perennial, 2014), 63.

16. The tribe limited egg collection to 10 percent of the estimated spawn for that season. Holtgren, "Bringing Us Back to the River," 141.

17. Other sturgeon restoration sites followed the LRBOI lead on streamside rearing facilities, including the Cedar, Whitefish, Kalamazoo, and Ontonagon Rivers in Michigan, and Wisconsin's Kewaunee and Milwaukee Rivers. US Fish and Wildlife Service, "Lake Sturgeon Streamside Rearing: Lake Michigan Partnership" (agency poster, n.d.).

18. J. Marty Holtgren et al., "Design of a Portable Streamside Rearing Facility for Lake Sturgeon, *North American Journal of Aquaculture* 69, no. 4 (2007): 317–323; Holtgren, "Bringing Us Back to the River," 139. This would also sustain the unique genetic makeup of this remnant population of Manistee River sturgeon. For an enthusiastic report on the first released sturgeon seen returning to the Manistee River system, and also "the first documented lake sturgeon from a Streamside Rearing Facility to return to its natal stream within the Great Lakes," see Matthew L. M. Fletcher, "Little River Band Ottawa Lake Sturgeon News," *Turtle Talk*, April 17, 2018. *Turtle Talk* is an Indigenous Law and Policy Center blog from the Michigan State University College of Law.

19. Marty Holtgren, Stephanie Ogren, and Kyle Whyte, "Renewing Relatives: One Tribe's Efforts to Bring Back an Ancient Fish," *Earth Island Journal* 30, no. 3 (Autumn 2015); Kyle Whyte, "What Do Indigenous Knowledges Do for Indigenous Peoples?" *Traditional Ecological Knowledge: Learning from Indigenous Practices for Environmental Sustainability*, ed. Melissa K. Nelson and Daniel Shilling (New York: Cambridge University Press, 2018), 71–72. See also Mitchell, "N'me," 21–25. Note that Holtgren and Ogren are the two tribal biologists with whom Gregory A. D. worked during filming.

20. Marty Holtgren, Stephanie Ogren, and Kyle Whyte, "Renewing Relatives: One Tribe's Efforts to Bring Back an Ancient Fish," *Earth Island Journal* 30, no. 3 (Autumn 2015). The envisioned connection also encompassed non-Indigenous residents, "settler Americans," as the LRBOI called them.

21. Holtgren, "Bringing Us Back to the River," 144.

22. MNA Staff, "Sturgeon Movie Premiere Monday at Ramsdell," *Manistee News Advocate* (Manistee, MI), November 2, 2011.

RIVER PEOPLE

1. Greg's "crew" were the shore anglers keeping an eye out for him.

2. I saw a version of this for myself, even though it was a cold early December day.

3. Laura Parker, "The Great Pacific Garbage Patch Isn't What You Think It Is: It's Not All Bottles and Straws; The Patch Is Mostly Abandoned Fishing Gear," *National Geographic*, March 22, 2018. For the original scientific study, see L. Lebreton et al., "Evidence that the Great Pacific Garbage Patch Is Rapidly Accumulating Plastic," *Scientific Reports* 8, no. 4666 (2018), https://doi.org/10.1038/s41598-018-22939-w.

4. David M. Chavis and Kien Lee, "What Is Community Anyway?" *Stanford Social Innovation Review* (Stanford, CA), May 12, 2015.

5. Kathleen M. MacQueen et al., "What Is Community? An Evidence-Based Definition for Participatory Public Health," *American Journal of Public Health* 91, no. 12 (December 1, 2001): 1929–1938.

6. MacQueen et al., "What Is Community?," 1930.

7. MacQueen et al., "What Is Community?," 1930. From their analysis, MacQueen et al. arrived at the following definition of community: "A group of people with diverse characteristics who are linked by social ties, share common perspectives, and engage in joint action in geographical locations or settings."

8. MacQueen et al., "What Is Community?," 1930–1932. The five core elements of community that MacQueen et al. identified and elaborated on were locus, sharing, joint action, social ties, and diversity. Sharing and diversity did not transcend all the study groups.

9. Most recent is Titz, Cannon, and Krüger's skeptical take on "community" when it comes to disaster relief and development initiatives, especially the potentially harmful assumption of a homogeneous "collective actor" on the ground, rather than heterogeneous groups with likely asymmetrical power relations both externally and internally. Alexandra Titz, Terry Cannon, and Fred Krüger, "Uncovering 'Community': Challenging and Elusive Concept in Development and Disaster Related Work," *Societies* 8, no. 3 (2018), https://doi.org/10.3390/soc8030071.

POWER IN THE VISUAL

1. Adéle Hast, "Domtar, Inc., History," *International Directory of Company Histories, vol. 4* (Farmington Hills, MI: St. James Press–Gale, 1991), 271–273. And here is Domtar's historical selfie: "Domtar: A History of Agility, Innovation and Caring," January 26, 2017, https://newsroom.domtar.com/history-domtar-corporation.

2. The name "Dominion" comes from "Dominion of Canada," referencing Canada's status as a dominion of the British Empire.

3. Thomas Derdak and Jay P. Pederson, eds., "Pentair, Inc., History," *International Directory of Company Histories, vol. 26* (Farmington Hills, MI: St. James Press–Gale, 1999), 362.

4. Nancy Langston traces the early twentieth-century life cycle of the "dilution" model of water pollution control. She shows how emerging scientific understandings of chemical transport across continents and bioaccumulation up food chains first challenged and then debunked earlier intuitive ideas about dilution. See Nancy Langston, "Changing Understandings of Dosage and Dilution," in *Sustaining Lake Superior: An Extraordinary Lake in a Changing World* (New Haven, CT: Yale University Press, 2017), 63–66.

5. Palmer had undertaken a tour of commercial fishing operators and reported his results to commission president Herschel Whitaker, who had charged Palmer with "[collecting] statistics as to the condition of the fisheries along the south-west shore of Lake Huron and the eastern shore of Saginaw bay." *Ninth Biennial Report of the State Board of Fish Commissioners: From Dec. 1, 1888, to Oct. 1, 1890* (Lansing, MI, 1890), 13.

6. *Ninth Biennial Report of the State Board of Fish Commissioners*, 13. Historian Margaret Beattie Bogue also quotes a visiting official who encountered "a dark

river, foul smelling with sulphurated hydrogen coming from decay. It would seem impossible for anything to live in it." Margaret Beattie Bogue, *Fishing the Great Lakes: An Environmental History, 1783–1933* (Madison: University of Wisconsin Press, 2000), 127.

7. Jim Lynch, "Port Huron Domtar Paper Plant under Scrutiny: State, Anglers Worry Discharge Is Polluting the St. Clair River," *Detroit News*, October 11, 2010.

8. Chris Dombrowski, *Body of Water: A Sage, a Seeker, and the World's Most Alluring Fish* (Minneapolis, MN: Milkweed, 2017), 179.

9. Dombrowski, *Body of Water*, 181.

10. Note that the Domtar controversy fell within a brief period of change to both the names and organizational structures of Michigan's principle environmental agencies. In 2009 via Executive Order 2009-45, Michigan governor Jennifer Granholm merged the Michigan Department of Environmental Quality (MDEQ) and the Michigan Department of Natural Resources (MDNR) into the Michigan Department of Natural Resources and Environment (MDNRE). In 2011 via Executive Order 2011-1, Michigan governor Rick Snyder reversed Granholm's merger, abolishing the MDNRE and re-establishing separate Departments of Natural Resources and Environmental Quality.

11. Domtar's MDNRE permit allowed for a daily average discharge of 3,308 pounds of suspended solids. Separately, Domtar Corporation made local water conservation a point of pride. "After water is used to make paper or to cool equipment, it is treated and then discharged into public waterways," the company said in its 2007 annual review. "To protect the population and the natural environment, strict environmental regulations have been implemented at the federal, state and provincial levels for what external pollutants this water can safely contain. Three of the main components in pulp and paper mill effluent are biological oxygen demand (BOD), total suspended solids (TSS) and adsorbable [*sic*] organic halides (AOX). In 2007, all of Domtar's mills recorded BOD, TSS, and AOX levels well below all allowable limits, often making the water discharged cleaner than the water pumped into the mill." See *Domtar Annual Review*, vol. 1 (2007), 63.

12. International Joint Commission (IJC), *Revised Great Lakes Water Quality Agreement of 1978 as Amended by Protocol, Signed November 18, 1987*. The Agreement was amended again in 2012 and included this specific language "acknowledging the vital importance of the Great Lakes to the social and economic well-being of both countries, the close connection between quality of the Waters of the Great Lakes and the environment and human health, as well as the need to address the risks to human health posed by environmental degradation." IJC, *Appendix to the Protocol Amending the Agreement Between Canada and the United States of America on Great Lakes Water Quality, 1978, as Amended on October 16, 1983 and on November 18, 1987: Agreement Between Canada and the United States of America on Great Lakes Water Quality, 2012*, 3. For a map of US Great Lakes areas of concern (AOC), see EPA, "U.S. Great Lakes Areas of Concern," *EPA.gov*, last modified May 15, 2019, https://www.epa.gov/sites/production/files/2019-06/documents/aoc_map_b3 _text_002.pdf. For a map of Canadian and US Great Lakes areas of concern, see Government of Canada, "Great Lakes: Areas of Concern," *Canada.ca*, last modified

January 29, 2020, https://www.canada.ca/en/environment-climate-change/services
/great-lakes-protection/areas-concern.html.

13. IJC, "Article II: Purpose," *Revised Great Lakes Water Quality Agreement of 1978 as
Amended by Protocol, Signed November 18, 1987*, 4.

14. MDNRE, "Strategy for Delisting Michigan's Areas of Concerns," April 16, 2010, 5;
IJC, "Annex II: Remedial Action Plans and Lakewide Management Plans," *Revised
Great Lakes Water Quality Agreement of 1978 as Amended by Protocol, Signed
November 18, 1987*, 24.

15. Oversight of the St. Clair River AOC involved an agreement among the
Environmental Protection Agency, Environment Canada (as of 2015, Environment
and Climate Change Canada), the Michigan Department of Natural Resources (later,
the Michigan Department of Natural Resources and Environment, then back to the
original), and the Ontario Ministry of Environment (later, the Ontario Ministry of
Environment and Climate Change).

A NOT-SO-OBJECTIVE INTRODUCTION TO THE
FISH CONSUMPTION ADVISORY

1. Dow opened the first plant in 1949, and the second in 1965. Ontario Water
Resources Commission, "Report on the Mercury Pollution of the St. Clair River
System" (Toronto: Ontario Ministry of the Environment, 1970).

2. For historical and technical overviews, see Environment and Climate Change
Canada, "Section 1: Introduction," *EPS-1=HA-2 Compliance with Chlor-Akali
Mercury Regulations, 1986–1989: Status Report*, http://www.ec.gc.ca/lcpe-cepa
/default.asp?lang=En&n=E7E0E329-1&offset=4&toc=show (web-archived).

3. Environment and Climate Change Canada, "Section 1: Introduction," 12–14.

4. John H. Hartig, "Lake St. Clair: Since the Mercury Crisis," *Water Spectrum* 15, vol.
1 (1983): 18–25. For a succinct overview of the crisis and later fish monitoring, see
Al Hayton, "Indicator: Mercury in Lake St. Clair Walleye," Ontario Ministry of the
Environment, http://cronus.uwindsor.ca/units/stateofthestraight/softs.nsf
/54ef3e94e5fe816e85256d6e0063d208/4506079a8efac1ca852573a3005000fa
/$FILE/hg-le-walleye.pdf. Nor were the repercussions limited to this most
industrialized of Great Lakes waterways. In her important Lake Superior study,
environmental historian Nancy Langston documents the cumulative impacts of
pulp and paper production on more remote Indigenous fishing communities,
including two decades (22,000 pounds) of mercury contamination at Grassy
Narrows, "one of the worst poisoning episodes associated with pulp and paper
production." Nancy Langston, *Sustaining Lake Superior: An Extraordinary Lake in a
Changing World* (New Haven, CT: Yale University Press, 2017), 77–78, 201.

5. See the stellar long-term participatory research of medical anthropologist
Christianne Stephens and her Walpole Island First Nation collaborators:
Christianne V. Stephens, "Toxic Talk at Walpole Island First Nation: Narratives
of Pollution, Loss and Resistance" (dissertation, McMaster University, 2009);
Christianne V. Stephens and Regna Darnell, "Keepers of the Water, Keepers
of the Fire: Building Bridges between Academic and Indigenous Knowledges

in Environmental Health Research," *International Journal of the Humanities* 5, no. 10 (2008), 105–114; Christianne V. Stephens and Regna Darnell, "Assessing Environmental Health Risks through Collaborative Research and Oral Histories: The Water Quality Issue at Walpole Island," in *The Nature of Empires and the Empires of Nature: Indigenous Peoples and the Great Lakes Environment*, ed. Karl Hele (Waterloo, Ontario: Wilfrid Laurier Press, 2013): 189–220; D. Jacobs et al., "Health Risk of the Walpole Island First Nation Community from Exposure to Environmental Contaminants: A Community-Based Participatory Research Partnership," in *Aboriginal Policy Research Volume IX: Health and Well-Being*, ed. Jerry B. White et al. (Toronto: Thompson Educational, 2013), 49–55.

6. Beth LeBlanc, "The 'Blob' Is Gone: Historic Spill Changed the Way We View the St. Clair River," *Lansing State Journal*, last modified October 13, 2014.

7. LeBlanc, "The 'Blob' Is Gone."

8. For the dysfunctional but enduring "phone tree" to notify authorities and communities of toxic spills into the St. Clair River, see Chad Selweski, "Toxic Chemical Spill? Wait for Our Fax," *Bridge Magazine*, October 31, 2017.

9. Stephens, "Toxic Talk at Walpole Island First Nation," 5, 152.

10. PCBs were banned under the Toxic Substance Control Act (TSCA) of 1976 (PL 94-469). From 1929 until 1979, PCBs had a wide array of industrial uses—transformers, capacitors, electrical equipment, insulation, adhesives, and so on. The scientific, agency, and environmentalist literature on PCBs is huge. A good starting point is EPA, "Polychlorinated Biphenyls (PCBs)," *Epa.gov*, last modified December 8, 2020, https://www.epa.gov/pcbs.

11. This last, on the mobility of PCBs, makes them an atmosphere-surface exchangeable pollutant, or ASEP. Judith A. Perlinger et al., "Measurements and Modeling Atmosphere-Surface Exchangeable Pollutants (ASEPs) to Understand Their Environmental Cycling and Planetary Boundaries," *Environmental Science and Technology* 50 (2016): 8932–8934.

12. For example, NOAA, "PCBs: Why Are Banned Chemicals Still Hurting the Environment Today?" Office of Response and Restoration, last modified December 20, 2020, https://response.restoration.noaa.gov/about/media/pcbs-why-are-banned-chemicals-still-hurting-environment-today.html.

13. "St. Clair River Area of Concern Canadian Section Status of Beneficial Use Impairments," (Ontario: Environment Canada and the Ontario Ministry of the Environment, September 2010), 6. Also, "St. Clair River Area of Concern Revised Canadian Delisting Criteria, 2010," (Ontario: CRIC Delisting Working Group, October 2010), 8–12. The 2006 US delisting criteria and 2010 Canadian delisting criteria were worded differently for the restrictions on fish and wildlife consumption beneficial use impairment (BUI). But they used the same criteria: removal of the impairment would occur when St. Clair River fish advisories and fish contaminant loads were the same as those in the larger Great Lakes (i.e., in non-AOC sites). Each listed chemical has its own Great Lakes industrial history, its own pathway into fish tissue. Mirex, for example, was a flame retardant, and also an insecticide in the unsuccessful government "war" to eradicate the non-native fire ant in the Southeast (a fiasco chronicled by Rachel Carson in her landmark book, *Silent*

Spring, which focused on Mirex's fellow chlorinated hydrocarbon, the insecticide heptachlor). Hooker Chemical Company manufactured Mirex in Niagara Falls, New York, from 1957 until 1975. The EPA prohibited its use in 1976. Mirex entered the Great Lakes via Lake Ontario, and that lake remained the most heavily impacted by the chemical. But Mirex has also declined to such an extent that one research team expects Mirex contaminant loads in fish to be below the risk threshold of consumption advisories. Finally, note the intertwining histories, as Hooker Chemical was also at the center of the Love Canal tragedy. Rachel Carson, *Silent Spring* (Boston: Houghton Mifflin, 1962), 161–172. Ruth Stringer and Paul Johnston, "Chlorinated Pesticides [Mirex]," *Chlorine and the Environment: An Overview of the Chlorine Industry* (New York: Springer, 2001): 249–251; Nilima Gandhi et al., "Is Mirex Still a Contaminant of Concern for the North American Great Lakes?" *Journal of Great Lakes Research* 41 (2015), 1114–1122.

14. Renamed the "Eat Safe Fish Guide," the most recent 2018 Michigan advisory continues to recommend no more than six meals per year for walleye in the St. Clair River, Lake St. Clair, and Lake Erie. "The Michigan Family Fish Consumption Guide" (Lansing: Michigan Department of Community Health, 2008); "The Ontario Guide to Eating Sport Fish" (Ontario Ministry of Environment, 2009); "Eat Safe Fish Guide: Southeast Michigan 2018" (Lansing: Michigan Department of Health and Human Services, 2018).

15. Harold H. Humphrey and John L. Hesse, *State of Michigan Sport Caught Fish Consumption Advisories: Philosophy, Procedures and Process, Draft Procedural Statement* (Lansing, MI: Department of Public Health, November 1986), 54; Stratus Consulting, "Michigan Fish Consumption Advisories," in *Fish Consumption Advisories in the Lower Fox River/Green Bay Assessment Area*, final report prepared for the US Fish and Wildlife Service, November 24, 1998, 4-1–4-2 (59–60).

16. Valoree S. Gagnon, Hugh S. Gorman, and Emma S. Norman, "Eliminating the Need for Fish Consumption Advisories in the Great Lakes Region: A Policy Brief," Michigan Technological University Great Lakes Research Center Contribution No. 50, March 7, 2018, 3.

17. Langston, "Fish Advisories," in *Sustaining Lake Superior*, 176; Sarah Coefield, "Lake St. Clair Fish Advisories Here to Stay," *Great Lakes Echo*, February 17, 2010.

18. For the Michigan Fish Consumption Advisory Program policy document explaining the program, its methods and metrics, and guidelines for fish advisories, see "Michigan Fish Consumption Advisory Program Guidance Document Version 4.0" (Lansing: State of Michigan, September 14, 2016).

19. "Michigan Fish Consumption Advisory Program Guidance Document Version 4.0," A-16.

20. For their insights on the deep implications of fish consumption advisories, I am indebted to Stephens, "Toxic Talk at Walpole Island First Nation"; Gagnon, Gorman, and Norman, "Eliminating the Need for Fish Consumption Advisories in the Great Lakes Region"; and T. Bruce Lauber, Nancy A. Connelly, and Jeff Niederdeppe, "What We Know about Fish Consumption Advisories: Insights from Experts and the Literature," *HDRU* 13, no. 6 (2013).

21. Lauber, Connelly, and Niederdeppe, "What We Know about Fish Consumption Advisories," A-17.

22. Florence Williams, "Toxic Breastmilk?" *New York Times*, January 9, 2005.

23. "Omega-3s: The Vanishing Nutrient," *In Defense of Food*, aired December 24, 2015, on PBS.

24. "Omega-3s: The Vanishing Nutrient"; Michael Pollan, *In Defense of Food: An Eater's Manifesto* (New York: Penguin, 2008).

25. "Fish: Friend or Foe?" *The Nutrition Source*, Harvard T. H. Chan School of Public Health, 2018, https://www.hsph.harvard.edu/nutritionsource/fish/.

26. In addressing the question, "Should Women Exposed to Environmental Toxins Breastfeed?," the Centers for Disease Control and Prevention say this: "Breastfeeding is still recommended despite the presence of chemical toxins. The toxicity of chemicals may be most dangerous during the prenatal period and the initiation of breastfeeding. However, for the vast majority of women, the benefits of breastfeeding appear to far outweigh the risks. To date, effects on the nursing infant have been seen only where the mother herself was clinically ill from a toxic exposure." "Breastfeeding: Environmental Exposures/Toxicants," *CDC.gov*, last modified June 23, 2020, https://www.cdc.gov/breastfeeding/breastfeeding-special -circumstances/environmental-exposures/index.html.

27. Langston's powerful, integrative history on Lake Superior ecology, toxic pollution, and Indigenous communities should be required reading for scholars and communities across the binational Great Lakes basin. Especially significant, to my mind, are the parallel experiences of Indigenous communities separated by great distances, whether along one of the most intensively industrialized stretches of the basin—the Huron–Erie corridor—or along one of the most seemingly pristine stretches—the Lake Superior shoreline of Langston's study. While the particulars will vary by location and community, their shared industrial histories also created shared, or at least comparable, regulatory and epidemiological histories—a severe pan-industrial experience, so to speak. See Langston, "The Mysteries of Toxaphene in Toxic Fish," in *Sustaining Lake Superior*, 165–185, and quotes from 176.

28. Or, as Jerry Dennis has observed, they might not believe the threat at all: "I've talked to dozens [of anglers] in every state and Ontario who admit frankly and even proudly that they ignore the advisories." Jerry Dennis, *The Living Great Lakes* (New York: Thomas Dunne/St. Martin's Press, 2003), 170.

29. See Kyle Powys Whyte, "Indigenous Experience, Environmental Justice, and Settler-Industrial States," in *Global Food, Global Justice: Essays on Eating Under Globalization*, ed. Mary C. Rawlinson and Cale Ward (Newcastle upon Tyne, UK: Cambridge Scholars, 2015), 143–156. "Food is associated with many hard-to-replace relationships," says Whyte in "Food Sovereignty, Justice, and Indigenous Peoples: An Essay on Settler Colonialism," in *Oxford Handbook of Food Ethics, ed.* A. Barnhill, T. Doggett, and A. Egan (Oxford, UK: Oxford University Press, 2018), 359. Gagnon, Gorman, and Norman note walleye harvesting as a tribal priority in upper Great Lakes states, and also in Great Lakes tribal communities "[consuming] fish several times a week." Gagon et al., "Eliminating the Need for Fish Consumption Advisories," 4–5.

30. The late Jim Harrison remains one of Michigan's and Montana's most famous writers. In addition to his poetry and novels (for example, *Legends of the Fall*), Harrison was a renowned food essayist. He celebrated "living to eat" in a column for *Esquire Magazine* called "The Raw and the Cooked." Jim Harrison, *A Really Big Lunch: The Roving Gourmand on Food and Life* (New York: Grove Press, 2017), 250–251.

31. Harrison, *A Really Big Lunch*, 2.

32. Harrison, *A Really Big Lunch*, 18–19.

33. Jim Harrison, "Cooking Your Life," in *The Raw and the Cooked: Adventures of a Roving Gourmand* (New York: Grove Press, 2001), 97.

34. Fram Dinshaw, "First Nations Lead Protest against Pollution in Ontario's Chemical Valley," *National Observer*, September 7, 2015.

35. Stephens, "Toxic Talk at Walpole Island First Nation," 179.

36. Sarah Marie Wiebe, *Everyday Exposure: Indigenous Mobilization and Environmental Justice in Canada's Chemical Valley* (Vancouver, British Columbia: UBC Press, 2016), 106–107.

37. Marc G. Weisskopf, Henry A. Anderson, and Lawrence P. Hanrahan, "Decreased Sex Ratio Following Maternal Exposure to Polychlorinated Biphenyls from Contaminated Great Lakes Sport-Caught Fish: A Retrospective Cohort Study," *Environmental Health* 2, no. 2 (2003), https://doi.org/10.1186/1476-069X-2-2.

38. According to Weisskopf, Anderson, and Hanrahan, "The odds of a male child among mothers in the highest PCB quintile compared to the lowest was reduced by 82%."

39. Weisskopf, Anderson, and Hanrahan, "Decreased Sex Ratio Following Maternal Exposure to Polychlorinated Biphenyls from Contaminated Great Lakes Sport-Caught Fish: A Retrospective Cohort Study."

40. Constanze A. Mackenzie, Ada Lockridge, and Margaret Keith, "Declining Sex Ratio in a First Nation Community," *Environmental Health Perspectives* 113, no. 10 (2005): 1295–1298. Note also, that as of 2001, the Aamjiwnaang First Nation had 850 members. The WIFN was larger. As of 2013, it had 4,521 members, with 2,277 of them living on the island itself.

41. Sarah Marie Wiebe, *Everyday Exposure: Indigenous Mobilization and Environmental Justice in Canada's Chemical Valley* (Vancouver, British Columbia: UBC Press, 2016), 143–145.

42. Wiebe, *Everyday Exposure*, 13, 88. For "slow violence," Wiebe cites R. Nixon, *Slow Violence and the Environmentalism of the Poor* (Cambridge, MA: Harvard University Press, 2011).

43. Wiebe details the "unflattering statistics" compiled by public health agencies. Wiebe, *Everyday Exposure*, 66–67.

44. Wiebe, *Everyday Exposure*, 29–30, 66–67.

45. Wiebe, *Everyday Exposure*, 103.

46. Wiebe, *Everyday Exposure*, 19.

47. See Wiebe's meticulous overview of the different scientific objections to the study, especially regarding scale, confounding variables, and the disadvantages to a small community of a scientific paradigm requiring large datasets at broad spatial scales.

Wiebe, *Everyday Exposure*, 145–146 and 151–154, citing Lambdon Community Health Study, "Reproductive Health," *Lambdon Community 2007 Health Status Report* (Point Edward, ON: Community Health Services Department, 2007).

48. Wiebe, *Everyday Exposure*, 153.
49. Wiebe, *Everyday Exposure*, 153, 176.
50. Stephens, "Toxic Talk at Walpole Island First Nation," 110.
51. David Treuer, *The Heartbeat of Wounded Knee: Native America from 1890 to the Present* (New York: Riverhead, 2019).
52. Emilie Cameron, "New Geographies of Story and Storytelling," *Progress in Human Geography* 36, no. 5 (2012): 572–591.
53. Cameron, "New Geographies of Story and Storytelling," 588.
54. Cameron, "New Geographies of Story and Storytelling," 581. I want to point readers to a book that will surely bring much more depth, insight, and sensitivity with its firsthand "storying" from Bkejwanong—Monty McGahey II's *Bkejwanong Dbaajmowinan/Stories of Where the Waters Divide* (East Lansing: Michigan State University Press, 2021).
55. Fram Dinshaw, "First Nations Lead Protest against Pollution in Ontario's Chemical Valley," *National Observer*, September 7, 2015.
56. Legal scholars Wenona Singel and Matthew Fletcher have long argued that American Indian and First Nation treaty rights are an essential component of saving Great Lakes waters and fisheries. See Wenona T. Singel and Matthew L. M. Fletcher, "Indian Treaties and the Survival of the Great Lakes," *Michigan State Law Review* 1285 (2006): 1285–1297. And as Gagnon, Gorman, and Norman explain and concur, "In the Great Lakes region, Anishinaabe nations negotiated treaties that explicitly reserve their hunting, fishing, and gathering rights across millions of acres in the basin. Contamination intrudes upon and erodes tribal harvesting practices protected by those treaties. For many US Native American tribes, fishing rights have been severely impacted by fish consumption advisories and toxicants." Gagnon, Gorman, and Norman, "Eliminating the Need for Fish Consumption Advisories in the Great Lakes Region," 7.
57. Marilyn Awama et al., "Two-Eyed Seeing and the Language of Healing in Community-Based Research," *Canadian Journal of Native Education* 32, no. 2 (2009): 3–23.
58. See Cindy Peltier, "An Application of Two-Eyed Seeing: Indigenous Research Methods with Participatory Action Research," *International Journal of Qualitative Methods* 17 (2018): 1–12.
59. Wiebe, *Everyday Exposure*, 106–114.
60. Wiebe, *Everyday Exposure*, 82–83.
61. Wiebe, *Everyday Exposure*, 141.
62. "The very thing that distinguishes Indigenous peoples from settler societies is their unbroken connection to ancestral homelands," says Dina Gilloo-Whitaker of the Colville Confederated Tribes. Dina Gillo-Whitaker, *As Long as Grass Grows: The Indigenous Fight for Environmental Justice from Colonization to Standing Rock* (Boston, MA: Beacon, 2019), 27.

1. Readers might also explore visionary restoration efforts in the Detroit River. See, for example, John H. Hartig, *Waterfront Porch: Reclaiming Detroit's Industrial Waterfront as a Gathering Place for All* (East Lansing: Greenstone, an imprint of Michigan State University Press, 2019); and *Bringing Conservation to Cities: Lessons from Building the Detroit River International Wildlife Refuge* (Burlington, ON: Aquatic Ecosystem Health and Management Society, 2014). For a sense of momentum in the St. Clair, the following beneficial use impairments (BUIs) were successfully removed later: Added cost to agriculture or industry (removed in 2011), tainting of fish and wildlife flavor (removed in 2012), degradation of aesthetics (removed in 2012), degradation of benthos (removed in 2014), bird or animal deformities or reproduction problems (removed in 2017), loss of fish and wildlife habitat (removed in 2017), and beach closings (removed in 2017). Two critical BUIs remain: restrictions on fish and wildlife consumption, and restrictions on drinking-water consumption. See EPA, "Restoring St. Clair River AOC (timeline), *EPA.gov*, last modified January 10, 2018, https://19january2017snapshot.epa.gov/st-clair-river -aoc/restoring-st-clair-river-aoc-timeline_.html.

2. Jim Lynch, "Port Huron Domtar paper plant under scrutiny: State, Anglers Worry Discharge Is Polluting the St. Clair River," *Detroit News*, October 11, 2010. Lynch's article reported on the Gregory A. D. video, the MDNRE response, and included interviews with anglers and conservation organizations.

3. *Domtar Annual Review*, vol. 1 (2007), 10.

4. *Domtar Annual Review*, vol. 1 (2007), 29.

5. Domtar Paper, "Crafting Paper Responsibly: Rooted in Sustainability from Forest to Print," *Domtar 2009 Sustainability Report*. This was Domtar's second sustainability report, and it covered 2007–2009.

6. Domtar Paper, "Crafting Paper Responsibly," 1.

7. Domtar Paper, "Crafting Paper Responsibly," 1.

8. Domtar was also certified for ISO 9001, and by other third-party organizations, including the Forest Stewardship Council, the Programme for the Endorsement of Forest Certification, the Sustainable Forestry Initiative, and Underwriters Laboratories.

9. Domtar Paper, "Crafting Paper Responsibly," 17.

10. Domtar Paper, "Crafting Paper Responsibly," 14.

11. Wartfroggy, "Domtar Paper Plant Pollution" thread in the Lake St. Clair and St. Clair River discussion forum, Michigan-Sportsman, October 12, 2010, https://www. michigan-sportsman.com/forum/threads/domtar-paper-plant-pollution.351767/.

12. Domtar Paper, "Crafting Paper Responsibly: Rooted in Sustainability from Forest to Print," *Domtar 2009 Sustainability Report*, Domtar Paper update, October 22, 2010.

13. Domtar Paper, "Crafting Paper Responsibly," Domtar Paper update, October 22, 2010.

14. Domtar Paper, "Corrective Action Update," December 8, 2010. The company submitted this report to the St. Clair River Bi-National Public Advisory Council (BPAC) meeting, Sarnia, Ontario, December 8, 2010.

15. Richard White, *The Organic Machine: The Remaking of the Columbia River* (New York: Hill and Wang, 1995).

16. The Federal Water Pollution Control Act amendments of 1972 established the NPDES under the jurisdiction of the Environmental Protection Agency (EPA). In Michigan, the EPA delegated administration of the NPDES to the state in 1973.

17. L. J. Van Dijk, R. Goldsweer, and H. J. Busscher, "Interfacial Free Energy as a Driving Force for Pellicle Formation in the Oral Cavity: An *In Vivo* Study in Beagle Dogs," *Biofouling* 1, no. 1 (1988): 19–25.

18. Greenwald is a Pulitzer-prize-winning journalist, best known for his national security reporting on the Edward Snowden/NSA documents. Glenn Greenwald and Leighton Akio Woodhouse, "Bred to Suffer: Inside the Barbaric U.S. Industry of Dog Experimentation," *The Intercept*, May 17, 2018.

19. T. D. Beckwith, "The Bacteriology of Pulp Slime," *Journal of Bacteriology* 22, no. 1 (1931): 15–22.

20. J. R. Sanborn, "Development and Control of Microorganisms in a Pulp and Paper Mill System," *Journal of Bacteriology* 26, no. 4 (1933): 373, 376.

21. Beckwith, "The Bacteriology of Pulp Slime," 20.

22. Heidi Annuk and Anthony P. Moran, "Microbial Biofilm-Related Polysachharides in Biofouling and Corrosion," in *Microbrial Glycobiology: Structures, Relevance and Applications*, ed. Anthony Moran et al. (Burlington, MA: Academic Press–Elsevier Science and Technology, 2009), 781, 791.

23. Hans-Curt Flemming, Michael Meier, and Tobias Schild, "Mini-Review: Microbial Problems in Paper Production," *Biofouling* 29, no. 6 (2013): 685.

24. Hans-Curt Flemming, "Microbial Biofouling: Unsolved Problems, Insufficient Approaches, and Possible Solutions," in *Biofilm Highlights*, ed. Hans-Curt Flemming, Ulrich Szewzyk, and Jost Wingender (Berlin: Springer-Verlag, 2011), 81–109. In another Flemming publication, "EPS: Then and Now," he traces a cultural history of slime, first to famous biologist and illustrator Ernst Haeckel's incorrect theory "that life originated from primordial slime," and Thomas Henry Huxley's equally mistaken belief that he had discovered Haeckel's primordial slime on the Atlantic sea floor. See Hans-Curt Flemming, "EPS: Then and Now," *Microorganisms* 4, no. 4 (2016), https://doi.org/10.3390/microorganisms4040041.

25. Flemming, "Microbial Biofouling," 82.

26. Hans-Curt Flemming et al. "Biofilms: An Emergent Form of Bacterial Life." *Nature Reviews Microbiology* 14 (2016): 563. Flemming advocates for an ecological perspective on biofilm. Theories within the discipline of ecology could provide important insights into biofilm microbiology and biochemistry, and conversely, ecology could derive important new insights from biofilm.

27. Mari Raulio, "Ultrastructure of Biofilms Formed by Bacteria from Industrial Processes" (dissertation, University of Helsinki, 2010), 28.

28. Much gratitude to ecologist Maarten Vonhof for clarifying the laboratory parameters of bacterial research.

29. In chronological order as identification keys improved: D. H. Eikelboom, "Filamentous Organisms Observed in Activated Sludge," *Water Research* 9, no. 4 (1975): 365–388; D. H. Eikelboom, "Identification of Filamentous Organisms

in Bulking Activated Sludge," *Progress in Water Technology* 8, no. 6 (1977): 153–161; D. H. Eikelboom and H. J. J. van Buijsen, "Identification of Filamentous Microorganisms," in *Microscopic Sludge Investigation Manual Report A94a*, 2nd ed. (Delft, Netherlands: TNO Research Institute for Environmental Hygiene, Water and Soil Division, 1983), 38–66; David Jenkins, Michael G. Richard, and Glen T. Daigger, *Manual on the Causes and Control of Activated Sludge Bulking and Foaming*, 2nd ed. (Boca Raton, FL: CRC Press–Taylor and Francis, 2003).

30. Anita Zumsteg, Simon K. Urwyler, and Joachim Glaubitz, "Characterizing Bacterial Communities in Paper Production: Troublemakers Revealed," *MicrobiologyOpen* 6, no. 4 (2017), https://doi.org/10.1002/mbo3.487.

31. Véronique Pellegrin et al., "Morphological and Biochemical Properties of a *Sphaerotilus* sp. Isolated From Paper Mill Slimes," *Applied and Environmental Microbiology* 65, no. 1 (1999): 156–162.

32. Zumsteg, Urwyler, and Glaubitz, "Characterizing Bacterial Communities in Paper Production." Both Zumsteg, Urwyler, and Glaubitz and Raulio also discuss research on *Deinococcus geothermalis*, which, describes Raulio, "operated as a pedestal for other bacteria [*Bacillus* sp.] to adhere and grow into a biofilm." See Mari Raulio, "Ultrastructure of Biofilms Formed by Bacteria from Industrial Processes" (dissertation, University of Helsinki, 2010), 34.

33. "There appear to be no typical strains associated with paper mills," according to Flemming, Meier, and Schild, "Mini-Review," 683.

34. Annuk and Moran, "Microbial Biofilm-Related Polysachharides in Biofouling and Corrosion," 782; Flemming, Meier, and Schild, "Mini-Review," 689; Water Environment Federation Task Force, "Filamentous Organisms," in *Wastewater Biology: The Microlife*, 2nd ed. (Alexandria, VA: Water Environment Federation, 2001), 93.

35. Flemming, Meier, and Schild, "Mini-Review," 68–-684. Also, Flemming, "Microbial Biofouling," 94–95.

36. Flemming et al. "Biofilms," 571–572.

37. Flemming, Meier, and Schild, "Mini-Review," 683–684.

38. Annuk and Moran, "Microbial Biofilm-Related Polysachharides in Biofouling and Corrosion," 790. On water recycling, see also Carolina Chiellini et al., "Bacterial Community Characterization in Paper Mill White Water," *BioResources* 9, no. 2 (2014): 2541–2559.

39. Flemming, Meier, and Schild, "Mini-Review," 683.

40. Rachel Carson, *Silent Spring* (Boston: Houghton Mifflin, 1962).

41. Flemming, Meier, and Schild, "Mini-Review," 692.

42. Flemming, "Microbial Biofouling," 98–99.

43. J. R. Sanborn, "Development and Control of Microorganisms in a Pulp and Paper Mill System," *Journal of Bacteriology* 26, no. 4 (1933): 378.

44. Flemming, "Microbial Biofouling," 102.

45. Domtar Port Huron, "Domtar Corrective Action Update" (in the author's possession), December 8, 2010.

46. Domtar Port Huron, "Domtar Corrective Action Update" (in the author's possession), December 8, 2010.

47. St. Clair River Bi-National Public Advisory Council Minutes (in author's possession) for September 14, 2016.

48. St. Clair River Bi-National Public Advisory Council Minutes (in author's possession) for June 9, 2011; January 14, 2014; March 25, 2014; June 10, 2014; October 21, 2014; January 26, 2016; March 30, 2016; July 13, 2016; September 14, 2016; and November 16, 2016.

49. St. Clair River Bi-National Public Advisory Council Minutes (in author's possession) for March 30, 2016; July 13, 2016; and September 14, 2016.

50. St. Clair River Bi-National Public Advisory Council Minutes (in author's possession) for December 9, 2014; January 26, 2016; July 13, 2016; and November 16, 2016.

A DAZZLING DISCOVERY

1. It is referred to as the Mount Clemens Fisheries Research Station in some publications.

2. For information on their study, I relied on the following: Michael V. Thomas and Robert C. Haas, "Capture of Lake Sturgeon with Setlines in the St. Clair River, Michigan," *North American Journal of Fisheries Management* 19 (1999): 610–612; Michael V. Thomas, "St. Clair Waterway, DNR," in *1999 Activities of the Central Great Lakes Bi-National Lake Sturgeon Group*, ed. Tracy D. Hill and Jerry R. McClain (paper presented at the Great Lakes Fishery Commission, Lake Huron Committee Meeting, Ann Arbor, Michigan, March 20–21, 2000, and Lake Erie Committee Meeting, Niagara-On-The-Lake, Ontario, March 29–30, 2000), 19–23; Michael V. Thomas and Robert C. Haas, "Abundance, Age Structure, and Spatial Distribution of Lake Sturgeon *Acipenser fulvescens* in the St. Clair System," Michigan Department of Natural Resources Fisheries Research Report 2076 (Harrison Township, MI: MDNR Lake St. Clair Fisheries Research Station, December 2004). The report also appeared as an article in the *Journal of Applied Ichthyology* 18 (2002): 495–501. Mike Thomas and Todd Wills, "History of the Lake St. Clair Fisheries Research Station, 1966–2003," Department of Natural Resources, Fisheries Division, Lake St. Clair Fisheries Research Station, last edited May 26, 2017, https://www.michigan.gov/documents/dnr/LSCFRS_history_thru_1979v2_387778_7.pdf.

3. Thomas and Haas, "Abundance, Age Structure, and Spatial Distribution of Lake Sturgeon *Acipenser fulvescens* in the St. Clair System," 2.

4. Thomas and Haas, "Abundance, Age Structure, and Spatial Distribution of Lake Sturgeon *Acipenser fulvescens* in the St. Clair System," 2.

5. E. F. Roseman et al., "Conservation and Management of Fisheries and Aquatic Communities in Great Lakes Connecting Channels," *Journal of Great Lakes Research* 40, Supplement 2 (2014): 1–6. St. Clair–Detroit River system and Huron–Erie Corridor are interchangeable terms. While sturgeon do return to their natal waters in the spring, they can nonetheless travel great distances during the year. Harkness and Dymond even pointed to the tagged "Sturgeon 41," found in Lake Michigan sometime before 1933. Where did 41 come from? Forty-one had been tagged twenty-eight years earlier in Lake St. Clair. See W. J. K. Harkness and J. R. Dymond, *The Lake Sturgeon: The History of Its Fishery and Problems of Conservation*

(Toronto: Department of Lands and Forests, Fish and Wildlife Branch, 1961), 19, 115, citing "Mystery of sturgeon '41' Solved," *The Fisherman* 2, no. 7 (1933), 12.

6. The best summary to date was probably Carole D. Goodyear et al., *Atlas of the Spawning and Nursery Areas of Great Lakes Fishes, vols. 6–8, St. Clair River; Lake St. Clair; Detroit River* (Washington, DC: US Fish and Wildlife Service, 1982). Lake sturgeon on pages 4–5, 4, and 3–4 respectively.

7. The Federal Endangered Species Act of 1973 classifies an endangered species as on the brink of extinction, and a threatened species as likely to reach the brink.

8. As Diana and Boase said, "This lack of knowledge stimulated Michigan DNR, University of Michigan, U.S. Fish and Wildlife Service–Alpena Fishery Resources Office, and the U.S. Geological Survey–Great Lakes Science Center to initiate complimentary studies on the sturgeon population." Jim S. Diana and Jim Boase, "St. Clair Waterway, University of Michigan School of Natural Resources and Environment," in *1999 Activities of the Central Great Lakes Bi-National Lake Sturgeon Group,* ed. Tracy D. Hill and Jerry R. McClain (paper presented at the Great Lakes Fishery Commission, Lake Huron Committee Meeting, Ann Arbor, Michigan, March 20–21, 2000, and Lake Erie Committee Meeting, Niagara-On-The-Lake, Ontario, March 29–30, 2000), 24.

9. Truedell operated the Lewis Truedell Fishery (Caseville, Michigan). Board of Fish Commissioners' statistical agent S. C. Palmer interviewed Truedell on February 8, 1890. Palmer interviewed commercial operators along the Huron–Erie corridor and Lake Huron Michigan coastlines. His central focus was whitefish, and opportunities for the state to propagate and stock whitefish. *Ninth Biennial Report of the State Board of Fish Commissioners: From Dec. 1, 1888, to Oct. 1, 1890* (Lansing, 1890), 17–18.

10. W. B. Scott and E. J. Crossman, *Freshwater Fishes of Canada,* Bulletin 184 (Ottawa, CA: Fisheries Research Board of Canada, 1973), 88.

11. Readers have a wealth of excellent literature to consult. I especially appreciate Harkness and Dymond, because Harkness entered fisheries research early enough (lake sturgeon in Lake Nipigon in 1922) that he met people who gave him first- or second-hand family accounts of their nineteenth-century experiences. See W. J. K. Harkness and J. R. Dymond, *The Lake Sturgeon: The History of Its Fishery and Problems of Conservation* (Toronto: Department of Lands and Forests, Fish and Wildlife Branch, 1961). Margaret Beattie Bogue provides a definitive *Fishing the Great Lakes: An Environmental History, 1783–1933* (Madison: University of Wisconsin Press, 2000), 158. For culture, ecology, and restoration, see essays in Nancy Auer and Dave Dempsey, eds., *The Great Lake Sturgeon* (East Lansing: Michigan State University Press, 2013). I benefited from James Philip Baker, *The Distribution, Ecology, and Management of the Lake Sturgeon (Acepenser Fulvescens Rafinesque) in Michigan* (master's thesis, University of Michigan, Ann Arbor, 1980), 4–12. A few key works and fisheries data compilations get cited in many if not most later articles and books. These include: Harkness and Dymond, *The Lake Sturgeon;* Scott and Crossman, *Freshwater Fishes of Canada,* 82–89; Wayne H. Tody, "Whitefish, Sturgeon, and the Early Michigan Commercial Fishery," in *Michigan Fisheries Centennial Report, 1873–1973* (Lansing: Michigan Department of Natural Resources, Fisheries Division,

1974), 45–60. Norman S. Baldwin et al., *Commercial Fish Production in the Great Lakes, 1867–1977*, technical Report no. 3 (Ann Arbor, Michigan: Great Lakes Fishery Commission, 1979). Regarding Great Lakes forest history and the white pine "Cutover," see Susan Flader, ed., *The Great Lakes Forest: An Environmental and Social History* (Minneapolis: University of Minnesota Press, 1983).

12. Harkness and Dymond, *The Lake Sturgeon*, 3.

13. Harkness and Dymond, *The Lake Sturgeon*, 4. William John Knox Harkness was a renowned fisheries scientist and fish biologist at the University of Toronto, and eventually chief of the Fish and Wildlife Branch of Canada's Department of Lands and Forests (1946–1960). John R. Dymond was also an important Canadian ichthyologist and a national leader in the Canadian conservation movement.

14. James W. Milner, "Report on the Fisheries of the Great Lakes; The Result of Inquiries Prosecuted in 1871 and 1872," in *United States Commission of Fish and Fisheries, Part II, Report of the Commissioner for 1872 and 1873* (1874), 74.

15. Tody, "Whitefish, Sturgeon, and the Early Michigan Commercial Fishery," 53. Scott and Crossman cover the same in *Freshwater Fishes of Canada*, 88.

16. Some would argue that the gray wolf was equally dishonored with wanton destruction. See Tody, "Whitefish, Sturgeon, and the Early Michigan Commercial Fishery," 51.

17. Tody, "Whitefish, Sturgeon, and the Early Michigan Commercial Fishery," 55.

18. Perhaps "the largest market for freshwater fish in the world," Sandusky began making caviar in 1855, and in 1860, processing smoked sturgeon. Harkness and Dymond, *The Lake Sturgeon*, 54, 67. See also the historical reports on Lake Erie lake sturgeon compiled in Carole D. Goodyear et al., *Atlas of the Spawning and Nursery Areas of Great Lakes Fishes, vol. 9, Lake Erie* (Washington, DC: US Fish and Wildlife Service, 1982), 5–7.

19. James W. Milner, "Report on the fisheries of the Great Lakes; the result of inquiries prosecuted in 1871 and 1872," in *United States Commission of Fish and Fisheries, Part II, Report of the Commissioner for 1872 and 1873* (1874), 72.

20. Milner, "Report on the Fisheries of the Great Lakes," 6. Milner contrasted Sandusky with Green Bay in a few places of his report. "At Green Bay, they are taken in great abundance, and are almost universally destroyed," Milner reproved. "They come into the nets in great numbers in early fall, and are pulled into the boats with the gaff-hook, and thrown upon the offal-heap" (10).

21. Said Milner, "Visiting a firm in Chicago, who handled smoked sturgeon, I learned that their books contained orders for much more than they could supply, and they were willing to pay a round price for the article." Milner, "Report on the Fisheries of the Great Lakes," 73. Nonetheless, three new technologies ramped up supply beyond anything imaginable earlier: pound nets, ruthlessly efficient at sweeping up all the fish in a location; the steam tug, so that fishermen could ply farther waters; and the power gill-net lifter, which, according to Wayne Tody, "enabled crews to handle a much larger quantity of nets, and it played an important part in maintaining the yield of an otherwise dwindling fishery." With this machinery of extraction, commercial fishers vacuumed up the fish of the Great Lakes. See Tody, "Whitefish, Sturgeon, and the Early Michigan Commercial Fishery," 46–49.

22. All the fish commissioners' quotes that follow come from "Sturgeon," in the *Ninth Biennial Report of the State Board of Fish Commissioners*, 44–46.

23. Scott and Crossman trace the word isinglass to "the middle Dutch word *huysenblasse*, meaning sturgeon bladder." Scott and Crossman, *Freshwater Fishes of Canada*, 88.

24. T. W. Bridge, "The Natural History of Isinglass," *Journal of the Institute of Brewing* 11 (1905): 509.

25. Bridge, "The Natural History of Isinglass," 508–531. The journal published Bridge's talk and attendees' discussion during the Institute of Brewing Midland Counties Section "Meeting Held at the Grand Hotel, Birmingham, on Thursday, February 18th, 1905."

26. Bridge, "The Natural History of Isinglass," 522.

27. Bridge, "The Natural History of Isinglass," 522.

28. A keg was 135 pounds of caviar. For trends in the prices of caviar and sturgeon between 1885 and 1957, see Harkness and Dymond, *The Lake Sturgeon*, 55–56, and citing W. S. Tower, "The Passing of the Sturgeon: A Case of the Un-paralleled Extermination of a Species," *Popular Science Monthly* 73 (1909): 361–371.

29. Bogue, *Fishing the Great Lakes*, 161.

30. Scott and Crossman, *Freshwater Fishes of Canada*, 88.

31. Great Lakes and Huron–Erie Corridor data originating from Baldwin et al., *Commercial Fish Production in the Great Lakes, 1867–1977*, and cited by others, including Baker, "*The Distribution, Ecology, and Management of the Lake Sturgeon (Acepenser Fulvescens Rafinesque) in Michigan*," 7–9, and Bogue, *Fishing the Great Lakes*, 160. State of Michigan sturgeon fisheries data from Tody, "Whitefish, Sturgeon, and the Early Michigan Commercial Fishery," 55. Dates differ slightly, with Baldwin et al. spanning 1879–1899 and Tody, 1880–1900. Also see the excellent graphs of historical sturgeon catches in Harkness and Dymond, *The Lake Sturgeon*, 68–71.

32. Dave Dempsey, "Sturgeon: The Great Lakes Buffalo," in *The Great Lake Sturgeon*, ed. Nancy Auer and Dave Dempsey (East Lansing: Michigan State University Press, 2013), 1–7.

33. An excellent compilation of lake sturgeon and habitat data by region and watershed is Center for Biological Diversity, "Petition [to the U.S. Fish and Wildlife Service] to List U.S. Populations of Lake Sturgeon (*Acipenser fulvescens*) as Endangered or Threatened under the Endangered Species Act," May 14, 2018, https://www.biologicaldiversity.org/species/fish/pdfs/Lake-Sturgeon-petition-5-14-18.pdf.

34. State of Michigan, *Journal of the Senate* 100 (October 24, 1973): 1657–1658; reprinted in *Michigan Fisheries Centennial Report, 1873–1973* (Lansing: Michigan Department of Natural Resources, Fisheries Division, 1974), 4.

35. Baker, "*The Distribution, Ecology, and Management of the Lake Sturgeon (Acepenser Fulvescens Rafinesque) in Michigan*," 26–27.

36. Tody, "Whitefish, Sturgeon, and the Early Michigan Commercial Fishery," 56.

37. Thomas and Wills, "History of the Lake St. Clair Fisheries Research Station, 1966–2003," 21.

38. Thomas and Wills, "History of the Lake St. Clair Fisheries Research Station, 1966–2003," 20–21.

39. Two years later the researchers also switched to a mesh trawl, in order to eliminate bycatches of smaller fish. See Thomas, "St. Clair Waterway, DNR," 20–21.

40. Thomas, "St. Clair Waterway, DNR," 20–21.

41. Beginning in 2001, they used PIT tags (passive integrated transponder tags) injected into a dorsal scute. Said Thomas, "these greatly improved the mark-recapture data by minimizing tag loss." Thomas and Wills, "History of the Lake St. Clair Fisheries Research Station, 1966–2003," 23.

42. Thomas and Haas, "Capture of Lake Sturgeon with Setlines in the St. Clair River, Michigan," 611.

43. Thomas and Haas, "Capture of Lake Sturgeon with Setlines in the St. Clair River, Michigan," 612.

44. Thomas and Haas, "Capture of Lake Sturgeon with Setlines in the St. Clair River, Michigan," 610–612.

45. Thomas, "St. Clair Waterway, DNR," 19.

46. Thomas and Haas, "Abundance, Age Structure, and Spatial Distribution of Lake Sturgeon *Acipenser fulvescens* in the St. Clair System," 14.

47. In 1999, the US Fish and Wildlife Service Alpena Fisheries Research Office (Alpena FRO) collaborated with the Earthwave Society on video production for the documentary, *Lake Sturgeon: Dinosaurs of the Great Lakes* (2000). The week of filming in the Huron–Erie corridor included sturgeon spawning at the cinder reef. See Tracy D. Hill and Jerry R. McClain, eds., *1999 Activities of the Central Great Lakes Bi-National Lake Sturgeon Group* (paper presented at the Great Lakes Fishery Commission, Lake Huron Committee Meeting, Ann Arbor, Michigan, March 20–21, 2000, and Lake Erie Committee Meeting, Niagara-On-The-Lake, Ontario, March 29–30, 2000), 17. Thomas and Haas described the Alpena RFO video system as "a custom built frame for stationary recording." Thomas and Haas, "Abundance, Age Structure, and Spatial Distribution of Lake Sturgeon *Acipenser fulvescens* in the St. Clair System," 14.

48. S. Jerrine Nichols, "St. Clair Waterway, USGS," *1999 Activities of the Central Great Lakes Bi-National Lake Sturgeon Group*, ed. Tracy D. Hill and Jerry R. McClain (paper presented at the Great Lakes Fishery Commission, Lake Huron Committee Meeting, Ann Arbor, Michigan, March 20–21, 2000, and Lake Erie Committee Meeting, Niagara-On-The-Lake, Ontario, March 29–30, 2000), 2.

49. Nichols, "St. Clair Waterway, USGS," 31.

50. Most of the examples here come from Thomas and Wills wonderfully detailed narrative chronology in Thomas and Wills, "History of the Lake St. Clair Fisheries Research Station, 1966–2003."

51. David J. Jude, Robert H. Reider, and Gerald R. Smith, "Establishment of Gobiidaae in the Great Lakes Basin," *Canadian Journal of Fisheries and Aquatic Sciences* 49, no. 2 (1992): 416–421.

52. D. J. Jude, J. Janssen, and G. Crawford, "Ecology, Distribution, and Impact of the Newly Introduced Round and Tubenose Gobies on the Biota of the St. Clair and Detroit Rivers," in *The Lake Huron Ecosystem: Ecology, Fisheries, and Management*, ed. M. Munawar, T. Edsall, and J. Leach (Amsterdam, Netherlands: Academic Publishing, 1995), 449.

53. Laurelyn Whitt, *Science, Colonialism, and Indigenous Peoples: The Cultural Politics of Law and Knowledge* (Cambridge, UK: Cambridge University Press, 2014). Ethnobotanist Janna Rose uses the term "scientific colonialism" in her accessible article, "Biopiracy: When Indigenous Knowledge Is Patented for Profit," *The Conversation,* March 7, 2016.

54. Anna Clark, *Poisoned City: Flint's Water and the American Urban Tragedy* (New York: Metropolitan Books, 2018); Mona Hanna-Attisha, *What the Eyes Don't See: A Story of Crisis, Resistance, and Hope in an American City* (New York: One World, 2018).

55. Marc Edwards, "A-List Actor but F-List Scientist: Mark Ruffalo Brings Fear and Misinformation to Flint" (blog post), in *Flint Water Study Updates,* ed. Siddthaartha Roy, May 16, 2016, http://flintwaterstudy.org/2016/05/a-list-actor-but-f-list-scientist-mark-ruffalo-brings-fear-and-misinformation-to-flint/.

56. Marc Edwards, "EXCLUSIVE! Mark Ruffalo's WATER DEFENSE Sampling Methods Revealed" (blog post), in *Flint Water Study Updates,* ed. Siddthaartha Roy, May 9, 2017, http://flintwaterstudy.org/2017/05/exclusive-mark-ruffalos-water-defense-sampling-methods-revealed.

57. Francis X. Donnelly, "From Hero to Pariah: Flint Water Expert Fights for His Reputation," *Detroit News,* April 25, 2019, updated April 26, 2019.

58. This holds true for my hometown Kalamazoo River, once the longest Superfund cleanup site in the United States, and later a victim of the biggest inland oil spill in American history when an Enbridge pipeline burst. Dr. Steve Hamilton, W. K. Kellogg Biological Station and Cary Institute for Ecosystem Studies, personal communication with the author.

59. Johannes Persson, Emma L. Johansson, and Lennart Olsson, "Harnessing Local Knowledge for Scientific Knowledge Production: Challenges and Pitfalls within Evidence-Based Sustainability Studies," *Ecology and Society* 23, no. 4 (2018): 38, http://dx.doi.org/10.5751/ES-10608-230438.

60. Elin Kelsey, "Integrating Multiple Knowledge Systems into Environmental Decision-Making: Two Case Studies of Participatory Biodiversity Initiatives in Canada and Their Implications for Conceptions of Education and Public Involvement," *Environmental Values* 12 (2003): 382. How might Kelsey's hierarchy reveal itself in Great Lakes fisheries research? Again, it's subtle, but consider the coin of any research realm: journal publications. In this realm, scientists are the central actors, naturally, because they carry out the research described in the articles. However, within those articles, the locals on whose local knowledge and field observations scientists rely, become reduced to nameless "anglers" or "divers." They receive this minimalist mention in methodology sections of journal articles. Once in a while, you'll see a named thanks in the acknowledgments. But acknowledgments are usually reserved for professional colleagues.

61. Kelsey, "Integrating Multiple Knowledge Systems into Environmental Decision-Making," 382. Also of interest is what Native American scholar Kyle Whyte calls the "supplemental value" of local knowledges to scientists—"the properties or behavior of particular plants and animals, ecosystem services, or local environmental change that scientists typically do not consider or have access to when they engage their

studies." Whyte's scope is immense: Indigenous knowledge systems. Kathy and Greg are not part of an Indigenous knowledge system, so they fall outside of Whyte's important insights about Indigenous knowledges, governance, and sustainability planning. But construed narrowly, Whyte's point about "access" applies, given Gregory A. D.'s unique access to and long-term experience in the St. Clair River. And finally, Persson, Johansson, and Olsson discuss ways to adapt "evidence-based" hierarchies of information to allow for the insights of local knowledge and practical experience. "Sustainability studies," they note, "are often, and for good reason, informed by practical experience . . . we need to draw on both scientific and local knowledge and understand local concerns." See Kyle Whyte, "What Do Indigenous Knowledges Do for Indigenous Peoples?" in *Traditional Ecological Knowledge: Learning from Indigenous Practices for Environmental Sustainability*, ed. in Melissa K. Nelson and Daniel Shilling (New York: Cambridge University Press, 2018), 62; and Johannes Persson, Emma L. Johansson, and Lennart Olsson, "Harnessing Local Knowledge for Scientific Knowledge Production: Challenges and Pitfalls within Evidence-Based Sustainability Studies," *Ecology and Society* 23, no. 4 (2018): 38.

62. The St. Clair River BPAC synopsizes subsequent analyses: "A complex interaction of ice, water, and earth has created a unique environment at the head of the St. Clair River producing sufficient depth to offer relative protection from freighter prop wash, constant high velocity water flow, and a supply of coarse grained substrates (Manny, 2011). . . . In fact, a number of St. Clair River fisheries researchers postulate that this area along Port Huron could be the principle sturgeon population source for the Central Great Lakes (Manny, Boase, personnel [*sic*] communication, 2011)." St. Clair River Bi-National Public Advisory Council Habitat Subcommittee, "Delisting Targets for Loss of Fish/Wildlife Habitat Beneficial Use Impairment of the St. Clair River Area of Concern," report submitted to the MDEQ Office of the Great Lakes, report adopted January 8, 2009, revised November 9, 2012, revised June 5, 2013, 13.

CURRENTS

1. The vignettes and dialogue in this chapter come out of in-depth and sometimes freewheeling discussions with Kathy Johnson and Greg Lashbrook about the ways in which they, as divers, know the St. Clair River. The dialogue is theirs; however, this is not an exact transcription. While my editing is not as aggressive as the incredible literary interviews in *the Paris Review*, I actively clustered material thematically and edited for clarity, flow, and specificity. This principally meant removing the filler words and verbal ticks people use when talking extemporaneously and informally (the proverbial "thinking out loud") unless they were important to Greg's or Kathy's emphasis, meaning, flow, or style of storytelling. When I omitted phrases or moved paragraphs ahead, I often did not use ellipses. My guiding principles were to retain Greg and Kathy's voices, stories, lessons, and interpretations and to help readers see this world below the surface of the St. Clair. (Also, for a helpful review of research on the literary interview as a genre, see Anneleen Masschelein et al., "The Literary Interview: Toward a Poetics of a Hybrid Genre," *Poetics Today* 35, nos. 1–2 (2014): 1–49.)

2. In this chapter, Kathy and Greg refer to three shipwrecks in the navigationally challenging St. Clair River rapids: three-mast wooden schooner *M. E. Tremble*, upbound (collision and sinking on September 8, 1890, one dead, carrying coal); two-mast wooden schooner-barge *Fontana*, downbound from Presque Isle, Michigan, to Cleveland, Ohio (collision and sinking on August 3, 1900, one dead, carrying 2,593 tons of iron ore); and three-mast wooden schooner-barge *John Martin*, downbound (collision and sinking September 21, 1900, as it tried to avoid the *Fontana* wreckage, four dead, carrying iron ore). For gripping newspaper accounts, see: "Sunk in a Collision: Propeller Wetmore Runs Down the Schooner Tremble," *Chicago Daily Tribune*, September 9, 1890, p. 6; "One Man Was Drowned: Big Schooner Fontana Sunk by the Schooner Santiago Yesterday," *Detroit Free Press*, Sunday, August 5, 1900, p. 6; "Complete Blockade: Yuma Sinks the Schooner Martin and Four Lives Were Lost," *Detroit Free Press*, September 22, 1900, p. 2. For history and specs on these three or other Great Lakes vessels, search Bowling Green State University's Great Lakes Vessels Online Index, http://greatlakes.bgsu.edu/vessel/view/002045.

3. Net sinkers (also called sinker stones or notched pebbles) are prehistoric stone artifacts that early fishing peoples probably used to weigh down and stabilize fishing nets made of the roots of spruce and willow. The stones Greg collects are flattish, often oval-shaped, with an indented "waist." Hunting for net sinkers in the wilder parts of the St. Clair is a high form of rock hounding for experienced divers.

(SEEING + KNOWING) × TIME = HOPE?

1. "Wildly curious" being a core value of the organization Cambridge Curiosity and Imagination.

2. Wilenski and Caroline Wendling, *Fantastical Guides for the Wildly Curious: Ways into Hinchingbrooke Country Park* (Huntington, England: Cambridge Curiosity and Imagination, 2013).

3. Robert Macfarlane, *Landmarks* (London, UK: Hamish Hamilton, 2015), 320.

4. Macfarlane, *Landmarks*, 315, 320.

5. Jack St. Rebor, "McElligot's Pool," *Seussblog*, November 18, 2012, https://seussblog.wordpress.com/2012/11/18/mcelligots-pool. On his website, St. Rebor also has a nice rundown of the books he used for research.

6. Quotes edited for length and clarity.

7. St. Clair River Bi-National Public Advisory Council Habitat Subcommittee, "Delisting Targets for Loss of Fish/Wildlife Habitat Beneficial Use Impairment of the St. Clair River Area of Concern," report submitted to the MDEQ Office of the Great Lakes, report adopted January 8, 2009, revised November 9, 2012, revised June 5, 2013. Greg found the unionid bed in June of 2012.

8. Patty Troy, "First St. Clair River Symposium Comes to the American Side of the River," news release, August 1, 2014.

9. Jim Bloch, "Written off as Doomed, Native Mussels Survive Zebra Mussel Invasion," *Voice News*, October 2, 2014.

10. US Fish & Wildlife Service, "America's Mussels: Silent Sentinels," *Midwest Region Endangered Species*, last updated May 29, 2019, https://www.fws.gov/midwest

/endangered/clams/mussels.html.

11. Matthew T. Rowe and David T. Zanatta, "Investigating the Genetic Variation and Structure of a Native Unionid Mussel in the Laurentian Great Lakes following an Invasion of Dreissenid Mussels," *Biological Invasions* 17 (2015): 351. See also David T. Zanatta et al., "Distribution of Native Mussel (Unionidae) Assemblages in Coastal Areas of Lake Erie, Lake St. Clair, and Connecting Channels, Twenty-five Years After a Dreissenid Invasion," *Northeastern Naturalist* 22, no. 1 (2015): 223–235.

12. The authors "found that the negative impact . . . may decrease about 10 years after initial zebra mussel colonization." Frances E. Lucy et al., "Zebra Mussel Impacts on Unionids: A Synthesis of Trends in North America and Europe," in *Quagga and Zebra Mussels: Biology, Impacts, and Control, Second Edition,* eds. Thomas F. Nalepa and Don W. Schloesser (Boca Raton, FL: CRC Press, 2014), 641.

13. Bloch, "Written off as Doomed, Native Mussels Survive Zebra Mussel Invasion."

NEGOTIATING ABUNDANCE AND SCARCITY

1. Sporhase v. Nebraska ex rel. Douglas, 458 U.S. 941 (1982).

2. US Army Corps of Engineers, "Six-State High Plains Ogallala Aquifer Regional Resources Study: Summary Report" (Washington, DC: US Army Corps of Engineers, Southwestern Division, 1982).

3. During his short-lived presidential campaign, New Mexico governor Bill Richardson (Democrat) proposed a national water policy and "dialogue between [western and eastern] states," asserting that "states like Wisconsin are awash in water." Dan Egan, "A Water Query from Out West: Hopeful's Interest in Great Lakes Renews Calls for Compact," *Milwaukee Wisconsin Journal Sentinel,* Oct. 6, 2007.

4. Peter Annin, *The Great Lakes Water Wars* (Washington, DC: Island, 2006, revised edition 2018); Dave Dempsey, *On the Brink: The Great Lakes in the 21st Century* (East Lansing: Michigan State University Press, 2004); Terence Kehoe, *Cleaning Up the Great Lakes: From Cooperation to Confrontation* (Dekalb: Northern Illinois University Press, 1997); Lee Botts and Paul Muldoon, *Evolution of the Great Lakes Water Quality Agreements* (East Lansing: Michigan State University Press, 2005).

5. Lynne Heasley, "Paradigms and Paradoxes of Abundance: The St. Lawrence River and the Great Lakes Basin" (paper presented at the First World Congress of Environmental History, Copenhagen, Denmark, August 2009).

6. James W. Feldman and Lynne Heasley, "Re-centering North American Environmental History," *Environmental History* 10, no. 3 (2007): 951–958. These stories of abundance involve subgenres of popular literature and academic scholarship, each with its own landscape types, regional memories, and cultural touchstones. On fisheries alone, readers might begin with Stephen Bocking, "Fishing the Inland Seas: Great Lakes Research, Fisheries Management, and Environmental Policy in Ontario," *Environmental History* 2 (1997): 52–73; Joseph E. Taylor III, *Making Salmon: An Environmental History of the Northwest Fisheries Crisis* (Seattle: University of Washington Press, 1999); Margaret Beattie Bogue, *Fishing the Great Lakes: An Environmental History, 1783–1933* (Madison: University of Wisconsin Press, 2000); and Michael J. Chiarappa and Kristin M. Szylvian, *Fish for All: An Oral*

History of Multiple Claims and Divided Sentiment on Lake Michigan (East Lansing: Michigan State University Press, 2003).

7. See William Cronon, "Landscapes of Abundance and Scarcity," in *The Oxford History of the American West*, ed. Clyde A. Milner II, Carol A. O'Connor, and Martha A. Sandweiss (New York: Oxford University Press, 1994), 603–37; and Martin Melosi, *Coping with Abundance: Energy and Environment in Industrial America* (Philadelphia: Temple University Press, 1985).

8. The foundational work of North American water history on which historians and theorists continue to build and elaborate is Donald Worster, *Rivers of Empire: Water, Aridity, and the Growth of the American West* (New York: Oxford University Press, 1992). For a contemporary theoretical overview of the social production of water scarcity and abundance in service to a modern hydraulic society, see Andrew Biro, "River-Adaptiveness in a Globalized World," in *Thinking with Water*, ed. Cecilia Chen, Janine MacLeod, and Astride Neimanis (Montreal: McGill-Queen's University Press, 2013), 166–184.

9. For a rich historiography of Powell in the arid American West, and especially his 1878 "Report on the Lands of the Arid Region of the United States," work backward from John Wesley Powell, *Seeing Things Whole: The Essential John Wesley Powell*, ed. William deBuys (Washington, DC: Island, 2001); Donald Worster, *A River Running West: The Life of John Wesley Powell* (New York: Oxford University Press, 2000); and Wallace Stegner, *Beyond the Hundredth Meridian: John Wesley Powell and the Opening of the West* (New York: Penguin, 1992).

10. Of the arid Canadian prairies, legal scholar David Percy says that "in a pattern that was familiar in the American West, the role played by water law in creating shortages became the subject of examination only after all efforts at augmenting the natural supply of water had been exhausted. In Canada, it became apparent only in the last two decades that *the basic model of prairie water law had never been designed to deal with water scarcity*." David Percy, "Responding to Water Scarcity in Western Canada," *Texas Law Review* 83, no. 7 (2005): 2097; emphasis is mine. See also Tristan M. Goodman, "The Development of Prairie Canada's Water Law, 1870–1940," in *Laws and Societies in the Canadian Prairie West, 1670–1940*, ed. Louis A. Knafla and Jonathan Swainger (Vancouver: UBC Press, 2005), 266–279. Jim Warren and Harry Diaz offer a sympathetic view of dryland farmers in *Defying Palliser: Stories of Resilience from the Driest Region of the Canadian Prairies* (Saskatchewan: University of Regina Press, 2012). Sterling Evans provides a terrific model for binational, transnational, and comparative scholarship on this border region in *Bound in Twine: The History and Ecology of the Henequen-Wheat Complex for Mexico and the American and Canadian Plains, 1880–1950* (College Station: Texas A&M Press, 2007). See also Christopher Armstrong, Matthew Evenden, and H. V. Nelles, *The River Returns: An Environmental History of the Bow* (Montreal: McGill-Queen's University Press, 2009); and Shannon Stunden Bower, *Wet Prairie: People, Land, and Water in Agricultural Manitoba* (Vancouver: UBC Press, 2011).

11. Norris Hundley, *Water in the West: The Colorado River Compact and the Politics of Water in the American West*, 2nd ed. (Berkeley: University of California Press, 2009).

12. Donald J. Pisani offers a sophisticated but manageable entry with *Water, Land, and Law in the West: The Limits of Public Policy, 1850–1920* (Lawrence: University Press of Kansas, 1996). Then, for an alternative economic and legal history of the prior appropriation doctrine, tackle David Schorr, *The Colorado Doctrine: Water Rights, Corporations, and Distributive Justice on the American Frontier* (New Haven, CT: Yale University Press, 2012). In Canada, the provinces of Alberta and British Columbia experimented with, but ultimately rejected, prior appropriation models.

13. Alicia Acuna and David Burke, "Colorado to California: Hands Off Our Water," *Fox News Politics*, January 28, 2015.

14. Marianne Goodland, "Rainbarrel Bill Dead for Session," *Colorado Statesman*, May 4, 2015.

15. A stunning example is Matt Black's portfolio in "The Dry Land," *New Yorker*, September 29, 2014, http://www.newyorker.com/project/portfolio/dry-land.

16. Jerry Dennis, *The Windward Shore: A Winter on the Great Lakes* (Ann Arbor: University of Michigan Press, 2012), 11–12.

17. Flint Water Advisory Task Force, *Final Report*, commissioned by the Office of Governor Rick Snyder, State of Michigan, March 21, 2016.

18. Or the third largest, if you exclude the Caspian Sea.

19. In a different region of the border itself, Hirt explores the unequal social consequences and the unsustainable ecological consequences of a bilateral "politics of abundance" running through the history of the Columbia and Fraser Rivers of the Pacific Northwest. Paul Hirt, "Developing a Plentiful Resource: Transboundary Rivers in the Pacific Northwest," in *Water, Place, and Equity: Tempering Efficiency with Justice*, ed. John M. Whiteley, Helen Ingram, and Richard Perry (Cambridge, MA: MIT Press, 2008), 147–188.

20. Donald J. Pisani, "Beyond the Hundredth Meridian: Nationalizing the History of Water in the United States," *Environmental History* 5, no. 4 (2000): 476. For a national environmental history of wetlands, and specifically wetland drainage, see Ann Vileisis, "Machines in the Wetland Gardens," in *Discovering the Unknown Landscape: A History of America's Wetlands* (Washington, DC: Island, 1997), 111–41. For an important water history of settlement and development forces and governmental policies that transformed water-abundant landscapes of the American South, see Craig Colten, *Southern Waters: The Limits to Abundance* (Baton Rouge: Louisiana State University Press, 2014), 41–115.

21. Willis F. Dunbar and George S. May, *Michigan: A History of the Wolverine State* (Grand Rapids, MI: William B. Eerdmans, 1995), 157.

22. Quoted in "Tisch: 'Drain Commissioner State's Most Powerful Man,'" *Argus Press*, May 17, 1979. Drain commissioners had and have the authority to condemn property, contract, assess all costs for drain work and projects to landowners in a designated drain, issue bonds, sue, and be sued—all unchecked by any other official or agency, accountable only to the laws of the Michigan Drain Code and the voters at election time. For a timeline of drain-related Michigan law, see Appendix 2-D in Michigan Department of Transportation and Tetra Tech MPS, *Drainage Manual*, January 2006, https://www.michigan.gov/stormwatermgt/0,1607,7-205--93193--,00 .html (see also the Michigan Drain Code of 1956). For a midwestern context of state

drainage district legislative history, see Mary R. McCorvie and Christopher L. Lant, "Drainage District Formation and the Loss of Midwestern Wetlands, 1850–1930," *Agricultural History* 67, no. 4 (1993): 13–39.

23. Philip Micklin, Nikolay Aladin, and Igor Plotnikov, eds., *The Aral Sea: The Devastation and Partial Rehabilitation of a Great Lake* (Berlin: Springer Earth System Sciences, 2013).

24. For an examination of the US interstate framework during that intense period of negotiation, see Noah D. Hall, "Toward a New Horizontal Federalism: Interstate Water Management in the Great Lakes Basin," *University of Colorado Law Review* 77 (2006): 405–56.

25. Dave Dempsey, *Great Lakes for Sale: From Whitecaps to Bottlecaps* (Ann Arbor: University of Michigan Press, 2008).

26. In the past decade, an important literature has emerged to rethink and theorize North American borderlands, with an emphasis on transnational and comparative history. See Benjamin Johnson and Andrew Graybill, eds., *Bridging National Borders in North America: Transnational and Comparative Histories* (Durham: Duke University Press, 2010); Michael Behiels and Reginald Stuart, eds., *Transnationalism: Canada–United States History into the Twenty-First Century* (Montreal: McGill-Queen's University Press, 2010); Matthew Evenden and Graeme Wynn, "Fifty Four, Forty, or Fight? Writing within and across Boundaries in North American Environmental History," in *Nature's End: History and the Environment,* ed. Sverker Sörlin and Paul Warde (New York: Palgrave Macmillan, 2009); and Victor Konrad and Heather Nicol, *Beyond Walls: Re-inventing the Canada–United States Borderlands* (New York: Ashgate, 2008). The scholarship also examines regionality at the border: Sterling Evans, *The Borderlands of the American and Canadian Wests: Essays on Regional History of the Forty-Ninth Parallel* (Lincoln: University of Nebraska Press, 2006); Kyle Conway and Timothy Pasch, eds., *Beyond the Border: Tensions across the 49th Parallel in the Great Plains and Prairies* (Montreal: McGill-Queen's University Press, 2013); John J. Bukowczyk et al., *Permeable Border: The Great Lakes Basin as Transnational Region, 1650–1990* (Pittsburgh: University of Pittsburgh Press, 2005); Ken Coates and John Findlay, eds., *Parallel Destinies: Canadian-American Relations West of the Rockies* (Seattle: University of Washington Press, 2002).

27. The Beaufort Sea is another binational marine zone in the Arctic North.

28. Feldman and Heasley, "Re-centering North American Environmental History." Steven C. High explores the economic decline of the region in *Industrial Sunset: The Making of the North American Rustbelt, 1969–1984* (University of Toronto Press, 2003).

29. Note that the Uruguay River makes up the entire (though shorter) border between Argentina and Uruguay.

30. Emma S. Norman, Alice Cohen, and Karen Bakker, eds., *Water without Borders? Canada, the United States and Shared Waters* (University of Toronto Press, 2013).

31. Boundary Waters Treaty, U.S.–Great Britain [for Canada], January 11, 1909, temp. State Dept. No. 548, 36 stat. 2448.

32. L. M. Bloomfield and Gerald F. Fitzgerald, *Boundary Waters Problems of Canada and the United States: The International Joint Commission, 1912–1958* (Toronto: Carswell, 1958); Murray Clamen and Daniel Macfarlane, "The International Joint Commission, Water Levels, and Transboundary Governance in the Great Lakes," *Review of Policy Research* 32, no. 1 (2015): 40–59.

33. See Daniel Macfarlane and Murray Clamen, eds., *The First Century of the International Joint Commission* (University of Calgary Press, 2020).

34. For example, on industrialized waters, the IJC offered seminal ideas on water policy, such as the "virtual elimination" of persistent toxic substances.

WATER, OIL, AND FISH

1. Richard White, *The Organic Machine: The Remaking of the Columbia River* (New York: Hill and Wang, 1995), 64.

2. Daniel Macfarlane, "Nature Empowered: Hydraulic Models and the Engineering of Niagara Falls," *Technology and Culture* 61, no. 1 (2020): 109–143; and Daniel Macfarlane, "'A Completely Artificial and Man-Made Cataract': The Transnational Manipulation of Niagara Falls," *Environmental History* 18, no. 4 (Oct. 2013): 759–84. On waterways as infrastructure, see also works such as Ashley Carse, *Beyond the Big Ditch: Politics, Ecology, and Infrastructure at the Panama Canal* (Cambridge, MA: MIT Press, 2014); Ashley Carse and Joshua A. Lewis, "Toward a Political Ecology of Infrastructure Standards: or, How to Think about Ships, Waterways, Sediment and Communities Together," *Environment and Planning A* 49, no.1 (2017): 9–28; Martin Reuss, "Rivers as Technological Systems," in *The Oxford Encyclopedia of the History of American Science, Medicine, and Technology*, vol. 2, ed. Hugh Richard Slotten (Oxford, UK: Oxford University Press, 2014).

3. On the history of the Chicago River, Chicago Diversion, and Chicago's general water history, see John W. Larson, *Those Army Engineers: A History of the Chicago District U.S. Army Corps of Engineers* (Chicago: US Army Corps of Engineers, Chicago District, 1979); David M. Solzman, *The Chicago River: An Illustrated History and Guide to the Rivers and Its Waterways* (University of Chicago Press, 2006); Richard Lanyon, *Building the Canal to Save Chicago* (Bloomington, IN: Xlibris, 2012); Harold Platt, *Sinking Chicago: Climate Change and the Remaking of a Flood-Prone Environment* (Philadelphia: Temple University Press, 2018); Joshua Salzmann, *Liquid Capital: Making the Chicago Waterfront* (Philadelphia: University of Pennsylvania Press, 2017); Libby Hill, *The Chicago River: A Natural and Unnatural History*, rev. ed. (Carbondale: Southern Illinois University Press, 2019); Richard Cahan and Michael Williams, *The Lost Panoramas: When Chicago Changed Its River and the Land Beyond* (Chicago: CityFiles Press, 2011); Matt Edgeworth and Jeff Benjamin, "What Is a River? The Chicago River as Hyperobject," in *Rivers of the Anthropocene*, ed. Jason M. Kelly, Philip Scarpino, Helen Berry, James Syvitski, and Michael Meybeck (Oakland: University of California Press, 2018). On energy use in Chicago, see Harold Platt, *The Electric City: Energy and the Growth of Chicago* (University of Chicago Press, 1991).

4. These withdrawals concern North America's two largest rivers, the Mississippi and the St. Lawrence.

5. "Top Ten Public Works Projects of the Century," American Public Works Association, http://www2.apwa.net/about/awards/toptencentury/default.htm.

6. The federal government initially set the Lake Michigan diversion at 4,167 cubic feet per second (cfs), but the allowable volume fluctuated.

7. Daniel Macfarlane, *Negotiating a River: Canada, the US, and the Creation of the St. Lawrence Seaway* (Vancouver: UBC Press, 2014), 35–39.

8. Hydroelectric production downstream was a secondary objective.

9. With a flow of 5,000 cfs, these are the largest engineered diversions into the Great Lakes.

10. Macfarlane, *Negotiating a River*.

11. Averaged out over a forty-year period.

12. Peter Annin, *The Great Lakes Water Wars* (Washington, DC: Island Press, 2006, revised edition 2018).

13. Arthur M. Woodward, *Charting the Inland Seas: A History of the U.S. Lake Survey* (Detroit: US Army Corps of Engineers, Detroit District, 1991).

14. The International Joint Commission (IJC) contends that the Chicago Diversion reduced the mean levels of Lakes Michigan and Huron by approximately 2.5 inches, Lake Erie by 1.7 inches, Lake Ontario by 1.2 inches, and Lake Superior by less than an inch. If the diversion reached its 10,000 cfs capacity, about three times its current rate, the impact on the lakes would be proportional. Thus, the early twentieth-century estimate of a half-foot drop is reasonable. International Joint Commission, *Report on Further Regulation of the Great Lakes* (Washington, DC, and Ottawa: International Joint Commission, 1976) (hereafter IJC 1976); International Joint Commission, *Report on Great Lakes Diversions and Consumptive Uses* (Washington, DC, and Ottawa: International Joint Commission, 1985) (hereafter IJC 1985).

15. See the special issue on Great Lakes–St. Lawrence as a system in *The Canadian Geographer* 60, no. 4 (Winter 2016).

16. IJC 1976; IJC 1985.

17. Daniel Macfarlane, *Fixing Niagara Falls: Environment, Energy, and Engineers at the World's Most Famous Waterfall* (Vancouver: UBC Press, 2020).

18. Murray Clamen and Daniel Macfarlane, "The International Joint Commission, Water Levels, and Transboundary Governance in the Great Lakes," *Review of Policy Research* 32, no. 1 (January 2015): 40–59.

19. Christopher Jones, *Routes of Power: Energy and Modern America* (Cambridge, MA: Harvard University Press, 2014).

20. Michelle Murphy, "Chemical Infrastructures of the St. Clair River," in *Toxicants, Health and Regulation Since 1945,* ed. Soraya Boudia and Nathalie Jas (London, UK: Pickering & Chatto, 2013), 104.

21. Even the permit for this was buried within "a Federal Register notice [that] flew under everyone's radar until the public feedback window expired." Garret Ellison, "Permit for 98-Year-Old Pipelines under St. Clair River Sparks Alarm," *MLive*, March 16, 2016.

22. Presumably named for its intricate navigation of Great Lakes states and waterways.

23. Joel Hood, "4 inspections of Romeoville Pipeline Found No Potential for Leaks, Company Says: Federal EPA Will Check to See if Any Risks Remain," *Chicago Tribune*, September 16, 2010.

24. The price increased by sixteen cents in the Chicago area, by ten cents farther out.

25. Joel Hood, "Hiccup in Romeoville, Indigestion at Pump, Gas Prices Climb in Wake of Pipeline Shutdown," *Chicago Tribune*, September 15, 2010.

26. The BP Deepwater Horizon catastrophe occurred earlier in the year, and dwarfed even the Kalamazoo spill in terms of damage. For the Kalamazoo River spill, see National Transportation Safety Board, *Pipeline Accident Report: Enbridge Incorporated Hazardous Liquid Pipeline Rupture and Release Marshall, Michigan July 25, 2010*, Pipeline Accident Report NTSB/PAR-12/01 (Washington, DC: National Transportation Safety Board, 2012). See also Elizabeth McGowan and Lisa Song, "The Dilbit Disaster: Inside the Biggest Oil Spill You've Never Heard Of, Part 1," *Inside Climate News*, June 26, 2012.

27. This includes fines and cleanup costs.

28. Eric Anderson, NOAA–Great Lakes Environmental Research Laboratory, "Predicting Currents in the Straits of Mackinac," (Ann Arbor, MI: NOAA–Great Lakes Environmental Research Laboratory), https://www.glerl.noaa.gov/pubs/brochures/straits.pdf; David J. Schwab, *Statistical Analysis of Straits of Mackinac Line 5: Worst Case Spill Scenarios* (Ann Arbor: University of Michigan Water Center, March 2016).

29. In the 1960s, Italsider, Italy's largest steel manufacturer at that time (and one of the largest in Europe), made the coated steel pipe for the entire Lakehead system.

30. David Schwab, quoted in "Straits of Mackinac 'Worst Possible Place' for a Great Lakes Oil Spill, U-M Researcher Concludes," *Michigan News* (Ann Arbor), July 10, 2014.

31. Jonathan Oosting and Leonard N. Fleming, "Schuette: Close Line 5 Pipelines on 'Definite Timetable,'" *Detroit News*, June 29, 2017; Keith Matheny, "Straits Pipeline Report Raises Fears of Disaster: Enbridge Says Oil, Gas Lines Supported," *Detroit Free Press*, June 1, 2017; Dynamic Risk Assessment Systems, *Alternatives Analysis for the Straits Pipeline* (Calgary, Alberta: final report prepared for the State of Michigan), project no. SOM-2017–01, document no. SOM-2017–01-RPT-001, October 26, 2017.

32. Jeff Alexander and Beth Wallace, "Sunken Hazard: Aging Oil Pipelines beneath the Straits of Mackinac an Ever-Present Threat to the Great Lakes" (Ann Arbor, MI: National Wildlife Federation, 2012).

33. Jones, *Routes of Power*, 143–144.

34. In the Kalamazoo River, Enbridge didn't act for more than a day because it thought there was an air blockage, not a spill.

35. Testimony of David Lodge in *Asian Carp and the Great Lakes: Hearing before the Subcommittee on Water Resources and the Environment of the Committee on Transportation and Infrastructure*, 111th Cong., 1 (2010), 10. See also Andrew Reeves, *Overrun: Dispatches from the Asian Carp Crisis* (Toronto: ECW Press,

2019); and Dan Egan, *The Death and Life of the Great Lakes* (New York: W. W. Norton, 2017), 151–186.

36. P. J. Perea, "Asian Carp Invasion: Fish Farm Escapees Threaten Native River Fish Communities and Boaters as Well," *Outdoor Illinois* 10, no. 5 (May 2002): 8.

37. James L. Oberstar, quoted in *Asian Carp and the Great Lakes*, 13, 32.

38. National Invasive Species Act, H.R. 4283 (104th)/P.L. 104–332 (Oct. 26, 1996). This was a reauthorization and update of the Indigenous Aquatic Nuisance Prevention and Control Act of 1990.

39. Dameon Pesanti, "Smith-Root Finds Innovation in Science, Business," *The Columbian*, February 7, 2016.

40. David V. Smith, Fish repelling apparatus using a plurality of series connected pulse generators to produce an optimized electric field, US Patent 4750451, June 14, 1988. Smith-Root Inc. holds twenty-four patents, starting with Identification Tag Implanting Machine, US Patent 3369525, in Feb. 1968, up to Systems and Methods for Aquatic Electrical Barrier Desynchronization, US Patent 9468198, in Oct. 2016.

41. Smith-Root, Inc. [inventors David V. Smith and Lee Roy Carstensen], Electric fish barrier for water intakes at various depths, US Patent 6978734 B1, December 27, 2005.

42. US Army Corps of Engineers, Chicago District, *Dispersal Barrier Efficacy Study: Efficacy Study Interim Report IIA, Chicago Sanitary and Ship Canal Dispersal Barriers; Optimal Operating Parameters Laboratory Research and Safety Tests* (Chicago: Government Printing Office, 2011).

43. Joel Hood, "Carp Creeps into Lake Calumet: Discovery of Invasive Fish Triggers New Calls to Legal Action," *Chicago Tribune*, June 23, 2010; Michael Hawthorne, "Asian Carp Discovered Close to Lake Michigan as Trump Pushes Budget Cuts," *Chicago Tribune*, June 23, 2017.

44. William Cronon, *Nature's Metropolis* (New York: W. W. Norton, 1992).

45. Paul Sutter identified hybridity as a "defining tendency of recent scholarship in American environmental history"; Paul S. Sutter, "The World with Us: The State of American Environmental History," *Journal of American History* 100, no. 1 (2013): 94–119. Unfortunately we can't do justice to the robust envirotech historiography and recent work that intersect with this study. Mark Fiege, Dolly Jorgensen, Martin Melosi, Joy Parr, Sara Pritchard, Jeffrey Stine, Joel Tarr, and Richard White have been in the vanguard of scholars elucidating hybrid natures. For recent scholarship, see Lynne Heasley and Daniel Macfarlane, eds., *Border Flows: A Century of the Canadian-American Water Relationship* (University of Calgary Press, 2016); Martin Reuss and Stephen H. Cutcliffe, eds., *The Illusory Boundary: Environment and Technology in History* (Charlottesville: University of Virginia Press, 2010).

46. Macfarlane, "'A Completely Artificial and Man-Made Cataract.'"

47. Daniel Macfarlane, "Fluid Meanings: Hydro Tourism and the St. Lawrence and Niagara Megaprojects," *Histoire Sociale/Social History* 49, no. 99 (June 2016): 327–46.

This experimental chapter adapts the format of the *Harper's Index*, including its concise citations. The numbers in bold represent a specific entry's numerical position in the resource list. For each resource I've also recommended a book.

1. **Tree Sources**

RECOMMENDED: Douglas W. Tallamy, *Nature's Best Hope: A New Approach to Conservation that Starts in Your Yard* (Portland, Oregon: Timber Press, 2020).

1, 2. The Spruce (New York); **3.** Donald I. Dickman and Larry A. Leefers, *The Forests of Michigan (Ann Arbor: University of Michigan Press, 2018)*, 109; **4.** Michigan State University Extension (East Lansing, MI); **5.** Dickman and Leefers, *The Forests of Michigan*, 62–66; **6.** Lynne Heasley, Western Michigan University (Kalamazoo, MI.); **7.** Theodore J. Karamansky, *Deep Woods Frontier (Detroit: Wayne State University Press, 1989)*, 61–63; **8.** Agnes Mathilda Larson, *The White Pine Industry in Minnesota (Minneapolis: University of Minnesota Press, 2007)*, 380; **9.** Heasley; **10, 11.** Dickman and Leefers, *The Forests of Michigan*, 146, 169; **12.** Stephen Pyne, *Fire in America (Seattle: University of Washington Press, 1997)*, 199–218; **13.** Denise Gess and William Lutz, *Firestorm at Peshtigo (New York: Holt, 2002)*, 211–215; **14.** Wisconsin Historical Society (Madison); **15.** WWF–Australia (Sydney); **16.** *American Forestry*, December 1922; **17.** The Henry Ford Museum (Dearborn, MI); **18.** David Lee, *Chainsaws: A History* (Madeira Park, BC: Harbour, 2006), 79–82; **19.** Karamansky, *Deep Woods Frontier*, 252–253; **20.** National Oceanic and Atmospheric Administration (Washington, DC); **21.** James E. Meeker, Joan E. Elias, and John A. Heim, *Plants Used by the Great Lakes Ojibwa* (Odanah, WI: Great Lakes Indian Fish and Wildlife Commission, 1993); **22.** Michigan State University Extension (East Lansing, MI); **23.** Global Nonviolent Action Database, Swarthmore College (Swarthmore, PA); **24.** Nathalie Butt et al., "The Supply Chain of Violence," *Nature Sustainability* 2 (2019): 742–747; **25.** Joe Jackson, *The Thief at the End of the World* (London: Duckworth, 2009), 10, 153–216; **26.** Environmental Defense Fund (New York); **27.** Global Forest Atlas, Yale University (New Haven, CT); **28.** Food and Agriculture Organization of the United Nations (Rome); **29.** Liberty Vittert, "The Amazon Has Lost Ten Million Football Fields of Forest in a Decade," *Smithsonian Magazine*, January 1, 2020; **30.** USDA Forest Service (Washington, DC); **31.** Timothy J. Fahey, Cornell University (Ithaca, NY); **32.** Jerry Sullivan, *Hunting for Frogs on Elston* (University of Chicago Press, 2004), 212; **33.** Douglas Tallamy, University of Delaware (Newark, DE); **34.** USDA Forest Service (Washington, DC); **35.** Toiletpaperhistory.net; **36.** Barry Kudrowitz, University of Minnesota (Minneapolis); **37.** T. W. Crowther et al., "Mapping Tree Density at a Global Scale," *Nature* 525 (2015): 201–205; **38.** John Evelyn, *Sylva (London, 1670)*, b2; **39.** Atlas Obscura (Brooklyn, NY); **40.** Boreal Songbird Initiative (Seattle, WA).

2. **Salt Sources**

RECOMMENDED: Mark Kurlansky, *Salt: A World History* (New York: Penguin Books, 2003).

1. American Mines Services (Boulder, CO); 2. ThumbWind (Caseville, MI); 3. Family Farm Livestock (OH); 4. May Swenson, "Eclogue," *Paris Review* 10 (1955): 97; 5. James Feyrer, Dimitra Politi, and David N. Weil, "The Cognitive Effects of Micronutrient Deficiency," *Journal of the European Economic Association* 15, no. 2 (2017): 355–387; 6. Earth Sciences Museum, University of Waterloo (Waterloo, ON); 7. Ella Davies, "Earth's Saltiest Place Makes the Dead Sea Look Like Tapwater," *BBC Earth*, Aug. 10, 2016; 8. Charles W. Cook, *The Brine and Salt Deposits of Michigan (London: Forgotten, 2018)*, 5; 9. Michigan State University Extension (East Lansing, MI); 10. Michigan Department of Environmental Quality (Lansing, MI); 11. Cook, *The Brine and Salt Deposits of Michigan*; 12. Robb Gillespie, *Insights into the Michigan Basin (Boulder, CO: Geological Society of America, 2013)*, 9; 13. Lynne Heasley, Western Michigan University (Kalamazoo, MI); 14. Davies, "Earth's Saltiest Place Makes the Dead Sea Look Like Tapwater"; 15. SaltWorks (Woodinville, WA); 16. Earth Sciences Museum, University of Waterloo (Waterloo, ON); 17. Lee McDowell, *Mineral Nutrition History* (Sarasota, FL: First Edition Design, 2017), 49–50; 18. Brian Fagan, *The Intimate Bond* (New York: Bloomsbury, 2016), 190; 19. Earth Sciences Museum, University of Waterloo (Waterloo, ON); 20. Donald C. Barton and Roland B. Paxson, "The Spindletop Salt Dome and Oil Field Jefferson County, Texas," *AAPG Bulletin* 9, no. 3 (1925): 594–612; 21. Davies, "Earth's Saltiest Place Makes the Dead Sea Look Like Tapwater"; 22. Kurlansky, *Salt*, 25–26; 23. *Midland Daily News*, August 14, 2019; 24. Stephen Bertman, Western Michigan University (Kalamazoo, MI); 25. Merchant Research & Consulting (Birmingham, England); 26. SaltWorks (Woodinville, WA); 27. Cary Institute of Ecosystem Studies (Millbrook, NY); 28. Michigan Department of Natural Resources (Lansing, MI); 29. rad Plumer, "How America Got Addicted to Road Salt," *Vox*, Jan. 25, 2015; 30, 31. Noelle Bye, "Road Salt Use in Winter a Growing Problem, Scientists Say," *Daily Record*, Jan. 5, 2020; 32. Deirdre Lockwood, "Winter Road Salt May Corrode Plumbing and Contaminate Water for Nearby Well Owners," *Chemical & Engineering News*, Dec. 5, 2018; 33. Cary Institute of Ecosystem Studies (Millbrook, NY); 34. Tafline Laylin, "How Michigan's Flint River Came to Poison a City," *Guardian*, Jan. 18, 2016; 35. Cary Institute of Ecosystem Studies (Millbrook, NY); 36. Earth Institute, Columbia University (New York); 37. Scripps Institution of Oceanography, UC San Diego; 38. Davies, "Earth's Saltiest Place Makes the Dead Sea Look Like Tapwater"; 39. Manish Kumar, Tyler Culp, and Yuexiao Shen, "Water Desalination," *Frontiers of Engineering*, 55–61; 40. International Desalination Association (Topsfield, MA).

3.   **Iron Sources**

RECOMMENDED: Vaclav Smil, *Still the Iron Age: Iron and Steel in the Modern World* (Cambridge, MA: Butterworth-Heinemann, 2016).

1. Nathaniel L. Erb-Satullo, "The Innovation and Adoption of Iron in the Ancient Near East," *Journal of Archaeological Research* 27 (2019): 557–607; 2. USGS Mineral Commodity Summaries (Washington, DC); 3, 4. World Steel Association (Brussels); 5. J. D. Verhoeven, A. H. Pendray, and W. E. Dauksch, "The Key Role of Impurities in Ancient Damascus Steel Blades," *JOM* 59, no. 9 (1998): 58–64; 6. Frederick Betz, *Managing Innovation Competitive Advantage from Change*, 3rd

ed. (Hoboken, NJ: Wiley, 2011), 40; **7.** World Steel Association (Brussels); **8.** USGS Mineral Commodity Summaries (Washington, DC); **9.** Calumet Regional Archives, Indiana University–Northwest (Gary, IN); **10.** Vaclav Smil, *Still the Iron Age* (San Diego: Elsevier Science, 2016), 74; **11.** John P. Hoerr, *And the Wolf Finally Came* (University of Pittsburgh Press, 2014), 93; **12.** E. W. Davis, *Pioneering with Taconite* (St. Paul, MN: Minnesota Historical Society, 1964); **13.** Great Lakes Shipwreck Museum (Whitefish Point, MI); **14.** Jeffrey T. Manuel, "Mr. Taconite," *Technology and Culture* 54, no. 2 (2013): 337; **15.** Nancy Langston, "The Wisconsin Experiment," *Places*, April 2017; **16.** J. P. Casey, "Drone Specialist COPTRZ Promotes ELIOS 2 Drone for Use in Mines," *Mining Technology*, May 5, 2020; **17.** J. R. Owen et al., "Catastrophic Tailings Dam Failures and Disaster Risk Disclosure," *International Journal of Disaster Risk Reduction* 42 (2020); **18.** Steel Canvas (West Des Moines, IA); **19–23.** Jonathan Waldman, *Rust (New York: Simon and Schuster, 2015),* 2–3, 13–33, 101; **24.** Ken Burns, "Our Monuments Are Representations of Myth, Not Fact," *Washington Post*, June 23, 2020; **25.** Robert S. Starobin, *Industrial Slavery in the Old South* (London: Oxford University Press, 1971), 14–15; **26.** Cliff Brown and Terry Boswell, "Strikebreaking or Solidarity in the Great Steel Strike of 1919," *American Journal of Sociology* 100, no. 6 (1995): 1493; **27.** David LaVigne, "The 'Black Fellows' of the Mesabi Iron Range," *Journal of American Ethnic History* 36, no. 2 (2017): 15; **28.** National Park Service/National Register of Historic Places (St. Louis County, MN); **29.** US Bureau of Labor Statistics (Washington, DC); **30.** Warren C. Whatley, "African-American Strikebreaking from the Civil War to the New Deal," *Social Science History* 17, no. 4 (1993): 527; **31.** Earth Institute, Columbia University (New York); **32.** Smil, *Still the Iron Age*, 160; **33.** NGO Shipbreaking Platform (Brussels); **34.** USDC IN/ND case 2:18-cv-00127, filed 4/02/18; **35.** USDC/IN/ND case 2:19-cv-00473-PPS-JPK, filed 12/11/19; **36.** 2018 Tribal Wild Rice Task Force Report (Minnesota); **37.** Mightymac.org; **38.** Global Tall Building Database of the CTBUH (Chicago, IL); **39.** Smil, *Still the Iron Age*, 200; **40.** M. Jimmie Killingsworth, *Walt Whitman and the Earth* (Iowa City: University of Iowa Press, 2009), 75–84.

4. **Sand Sources**

RECOMMENDED: Vince Beiser, *The World in a Grain: The Story of Sand and How It Transformed Civilization* (New York: Riverhead Books, 2018).

  1. David Blatner, *Spectrums* (New York: Bloomsbury, 2014), 18–20; **2.** Carl Sagan, *Cosmos* (New York: Random House, 1980), 207; **3.** Yale Environment 360, Yale University (New Haven, CT); **4, 5.** United Nations Environment Programme (Nairobi); **6.** Vince Beiser, "Sand Mining," *Guardian*, Feb. 27, 2017; **7, 8.** UNEP Global Environmental Alert Service (Nairobi); **9.** Robert W. Kelly, *Geological Sketches of Michigan Sand Dunes* (Michigan Department of Conservation, 1962), 4; **10, 11.** Mary Ellen Benson, Anna B. Wilson, Donald I. Bleiwas, *Frac Sand in the United States* (Reston, VA: US Geological Survey, 2015), 5, 8; **12, 13.** Vince Beiser, "The Super-Secret Sand That Makes Your Phone Possible," *Wired*, August 7, 2018; **14.** IMARC Group (Noida, Uttar Pradesh); **15.** Mineral Commodity Summaries, USGS (Washington, DC); **16.** Lee Bergquist, "Sand Mining Company Fined," *Milwaukee Journal Sentinel*, Dec. 17, 2013; **17.** Benson, Wilson, and Bleiwas, *Frac Sand in the United States*, 53; **18.** Thomas P. Dolley, *2016 USGS Minerals Yearbook:*

Silica (US Geological Survey); **19.** Michigan DNR (Lansing, MI); **20.** Michigan Oil and Gas Association (Lansing, MI); **21, 22.** Kimberly Amadeo, *The Balance*, May 1, 2020; **23.** Bethany McLean, "America's Most Reckless Billionaire," *Saudi America* (New York: Columbia Group, 2018); **24.** *World Oil*, June 28, 2020; **25, 26.** Alison and David Swan, Saugatuck Dunes Coastal Alliance (Saugatuck, MI); **27, 28.** Tiffany Schriever, Western Michigan University (Kalamazoo, MI); **29.** Henry Chandler Cowles, *The Ecological Relations of the Vegetation of the Sand Dunes of Lake Michigan* (Chicago, 1899); **30, 31.** Alan F. Arbogast et al., *Valuing Michigan's Coastal Dunes* (Michigan Environmental Council, 2018), 71; **32, 33.** Dave LeMieux, "Preservationists Unable to Save Pigeon Hill from Sand Miners," *Muskegon Chronicle*, June 21, 2010; **34.** Jerry Dennis, *The Living Great Lakes* (New York: Thomas Dunne, 2003), 61; **35.** Michigan DEQ (Lansing, MI); **36.** Lake Michigan Federation (Chicago, IL, and Muskegon, MI); **37.** Dennis, *The Living Great Lakes*, 61; **38.** UNEP Global Environmental Alert Service (Nairobi); **39.** Michigan DEQ (Lansing, MI); **40.** ABC News, *Good Morning America* (New York)

THE PARADOX OF ABUNDANCE

1. John James Audubon, *Ornithological biography, or, An Account of the Habits of the Birds of the United States of America* (Edinburgh, UK, 1835), 319–327; "Plate 62 Passenger Pigeon," in *The Birds of America; from Original Drawings, vol. 1–4* (London, 1827–1838).

2. "Monarch of the Plains: The Great Slaughter of American Bison," *Nebraska State Journal* (Lincoln, Nebraska), September 7, 1887.

3. Donald Worster, *The Wealth of Nature: Environmental History and the Ecological Imagination* (New York: Oxford University Press, 1993).

4. Nick Squires, "Italy's Black Widow Refuses Day Release as She Does Not Want to Work," *Telegraph (London, UK)*, October 19, 2011.

5. Squires, "Italy's Black Widow Refuses Day Release as She Does Not Want to Work."

6. James Risen, "U.S. Identifies Mineral Riches in Afghanistan: Economic Boon Is Seen; Expanded Wealth Could Escalate Conflict and Corruption," *New York Times*, June 14, 2010; Stephen G. Peters et al., eds., and the US Geological Survey–Afghanistan Ministry of Mines Joint Mineral Resource Assessment Team, 2007, preliminary non-fuel mineral resource assessment of Afghanistan (US Geological Survey Open-File Report 2007–1214).

7. Jared Diamond, *Collapse: How Societies Choose to Fail or Succeed* (New York: Penguin, 2004).

8. Juliana Francis, "Community Living in Bondage of Gas Flaring," *Probe*, May 18, 2018.

9. Legborsi Saro Pyagbara, "The Adverse Impacts of Oil Pollution on the Environment and Wellbeing of a Local Indigenous Community: The Experience of the Ogoni People of Nigeria," United Nations Department of Economic and Social Affairs International Expert Group Meeting on Indigenous Peoples and Protection of the Environment, Khabarovsk, Russian Federation, August 27–29, 2007; Amarachi Paschaline Onyena and Kabari Sam, "A Review of the Threat of Oil Exploitation

to Mangrove Ecosystem: Insights from Niger Delta, Nigeria," *Global Ecology and Conservation* 22 (June 2020): https://doi.org/10.1016/j.gecco.2020.e00961.

10. Okechukwu Ibeanu, "Oiling the Friction: Environmental Conflict Management in the Niger Delta, Nigeria," *Environmental Change and Security Project Report* 6 (Summer 2000): 19–32.

11. Upton Sinclair, *Oil!* (New York: Grosset and Dunlap, 1927).

12. Donald Worster, "The Nature We Have Lost," in *The Wealth of Nature: Environmental History and the Ecological Imagination* (New York: Oxford University Press, 1993), 3–15.

13. C. Wright Mills, *The Power Elite* (New York: Oxford University Press, 1956).

14. Peter Annin, *The Great Lakes Water Wars* (Washington, DC: Island Press, 2006).

THE ACCIDENTAL REEF

1. S. Jerrine Nichols, "St. Clair Waterway, USGS," *1999 Activities of the Central Great Lakes Bi-National Lake Sturgeon Group*, ed. Tracy D. Hill and Jerry R. McClain (paper presented at the Great Lakes Fishery Commission, Lake Huron Committee Meeting, Ann Arbor, Michigan, March 20–21, 2000, and Lake Erie Committee Meeting, Niagara-On-The-Lake, Ontario, March 29–30, 2000), 33–34; also B. A. Manny and G. W. Kennedy, "Known Lake Sturgeon (*Acipenser fulvescens*) Spawning Habitat in the Channel between Lakes Huron and Erie in the Laurentian Great Lakes," *Journal of Applied Ichthyology* 18 (2002), 486–490.

2. Nichols, "St. Clair Waterway, USGS," 34.

3. Michael V. Thomas and Robert C. Haas, "Abundance, Age Structure, and Spatial Distribution of Lake Sturgeon *Acipenser fulvescens* in the St. Clair System," Michigan Department of Natural Resources Fisheries Research Report 2076 (Harrison Township, MI: MDNR Lake St. Clair Fisheries Research Station, December 2004),12.

4. Nichols, "St. Clair Waterway, USGS," 34.

Index

Port Huron Sulphite and Paper Company, 53–54

potamodromous species, 156 (n. 8)

pound nets, 79, 80, 187 (n. 21)

Powell, John Wesley, 110. *See also* Colorado River

Programme for the Endorsement of Forest Certification, 57, 182 (n. 8)

Public Advisory Council, 115; binational, 56, 57

public health, 51–52, 61, 63, 65

public trust doctrine, 114, 124–125

QMS (Quality Management System), 57

quagga mussels (*Dreissena rostriformis bugensis*): DNA database of, 163 (n. 11); effects of, 18, 161 (n. 7); food web and, 16; habitat of, 18; impact of, 161–162 (n. 10); as invasive species, 126; in Lake Huron, 18; in Lake Michigan, 18, 104–105; population growth of, 18; spread of, 126

Quality Management System, 57

Quebec, 53, 110

Rafinesque, Constantine Samuel, 24, 166 (n. 28), 170 (n. 62)

rainbow darters, 101

rainbow smelt, 82

RAP (Remedial Action Plan), 56, 57

Rashoman: as narrative mystery, 14; underwater, 15

Rashoman (film) (Kurosawa), 159 (n. 31); place as reef in, 14

"Rashomon effect" narrative, 14

Red Cedar River, 54

redhorse suckers, 83, 97; on accidental reef, 149, 150

reefs: accidental, 3, 5, 147, 149–50; artificial, 83, 84; coal cinders and, 3, 5, 83, 86, 149; coal clinkers and, 3, 7; documentary films of spawning on, 86, 189 (n. 47); Lake St. Clair, 3, 5, 15; lake sturgeon spawning and, 3–5, 82–83, 149–150; lake sturgeon survival at, 16; size of, 5; walleye on, 149

regional foods: Jim Harrison and, 62–63, 180 (n. 30); sustainability in, 62. *See also* lake whitefish; walleye; yellow perch

Remedial Action Plan, 56, 57

Remedial Action Plan Preparation and Adoption, 56

reproduction: choices, 44; health, 60; Indigenous birth and, 64, 66; pollution problems and, 17–18, 81, 182 (n. 1); strategy ("r"), 14; sturgeon, 81

research vessel (R/V), 81, 84

rhetoric, environmental, 14, 15, 68. *See also* metaphors; "Rashomon Effect" narrative

Richardson, Bill, 193 (n. 3)

riffleshell clam (*Epioblasma torulosa rangiana*), 42, 77, 149

riparian land, 111

Riskin, Jessica, 30

roe, 4. *See also* caviar

Romeoville, IL, 117, 122, 124, 126

Root, Lee, 127. *See also* Smith-Root Inc.

Rose, Janna, 190 (n. 53)

Rotterdam, Netherlands, 7

Rouge River Aesthetic Index, 57

round goby: avian botulism and, 18; as bait, 82; in Detroit River, 83; in Lake Michigan, 105; pollution and, 161 (n. 5); Ponto-Caspian roots of, 16, 18, 82; predators of, 18, 82; prey for, 16, 17, 18, 160 (n. 3); size of, 46; in St. Clair River, 16, 83, 84; at Traverse City, 105; zebra mussels and, 18, 160 (n. 3)

Roy, Siddhartha (Sid), 85

Royal Dutch Shell, 144–145

Royal Polymers Ltd., 59

RRAI (Rouge River Aesthetic Index), 57

"r" strategists, 14

Ruffalo, Mark, 85

runoff, 18, 161 (n. 7)

Russia: Aral Sea, 113, 146, 147; canals of, 11; trade of, 12

Russian Empire: Black Sea access and, 12, 13; canals as roadways in, 11, 12; Catherine II built world naval power of, 12, 13; caviar from, 80; Emperor Paul and canals in, 13; Peter the Great built canals in, 11, 12, 13; sturgeon air bladders in trade, 79–80; Vyshnii Volochek waterway in, 12–13; zebra mussels in canals of, 7, 11–13, 156 (n. 2)

R/V (research vessel), 81, 84

R/V Channel Cat (vessel), 81–82

Saginaw Bay, fisheries along, 174 (n. 5)

salmon: alewives and, 17, 18; chinook, 17; coho, 48, 163 (n. 4); jack, 163–164 (n. 12)

salt: brine, 5, 36, 137; Dow and, 5, 137; in *Harper's Index* format, 133, 136–137, 201; in Michigan, 132, 136; wells, 137

salt mines: brine deposits from, 5, 137; on Lake St. Clair, 5, 150; in Michigan, 132, 136

saltwater: desalinization and, 137; habitats, 156 (n. 8)

Sanborn, J. R., 72, 73, 75

sand: dune protection and, 141; Great-Lakes-coast, 140, 141; in *Harper's Index* format, 140; Michigan mining of, 132, 140, 141; in Saugatuck Dunes, 141; Sleeping Bear Dunes National Lakeshore, 18, 104, 141, 161 (n. 9); "Northern White," 140

Sander vitreus, 22, 33; Mitchill naming of, 23–24. *See also* walleye